Lecture Notes in Computer Science 15928

Founding Editors

Gerhard Goos
Juris Hartmanis

Editorial Board Members

Elisa Bertino, *Purdue University, West Lafayette, IN, USA*
Wen Gao, *Peking University, Beijing, China*
Bernhard Steffen ⓘ, *TU Dortmund University, Dortmund, Germany*
Moti Yung ⓘ, *Columbia University, New York, NY, USA*

The series Lecture Notes in Computer Science (LNCS), including its subseries Lecture Notes in Artificial Intelligence (LNAI) and Lecture Notes in Bioinformatics (LNBI), has established itself as a medium for the publication of new developments in computer science and information technology research, teaching, and education.

LNCS enjoys close cooperation with the computer science R & D community, the series counts many renowned academics among its volume editors and paper authors, and collaborates with prestigious societies. Its mission is to serve this international community by providing an invaluable service, mainly focused on the publication of conference and workshop proceedings and postproceedings. LNCS commenced publication in 1973.

Giacomo Bergami · Paul Ezhilchelvan ·
Yannis Manolopoulos · Sergio Ilarri ·
Jorge Bernardino · Carson K. Leung ·
Peter Z. Revesz
Editors

Database Engineered Applications

29th International Symposium, IDEAS 2025
Newcastle upon Tyne, UK, July 14–16, 2025
Proceedings

Editors
Giacomo Bergami
Newcastle University
Newcastle upon Tyne, UK

Paul Ezhilchelvan
Newcastle University
Newcastle upon Tyne, UK

Yannis Manolopoulos
Aristotle University of Thessaloniki
Thessaloniki, Greece

Sergio Ilarri
University of Zaragoza
Zaragoza, Spain

Jorge Bernardino
Institute of Education and Sciences (ISEC)
Lisbon, Portugal

Carson K. Leung
University of Manitoba
Winnipeg, MB, Canada

Peter Z. Revesz
University of Nebraska-Lincoln
Lincoln, NE, USA

ISSN 0302-9743 ISSN 1611-3349 (electronic)
Lecture Notes in Computer Science
ISBN 978-3-032-06743-2 ISBN 978-3-032-06744-9 (eBook)
https://doi.org/10.1007/978-3-032-06744-9

© The Editor(s) (if applicable) and The Author(s), under exclusive license to Springer Nature Switzerland AG 2026
Chapter "Data Mining for Language Superfamilies Using Congruent Sound Groups" is licensed under the terms of the Creative Commons Attribution 4.0 International License (http://creativecommons.org/licenses/by/4.0/). For further details see license information in the chapter.

This work is subject to copyright. All rights are solely and exclusively licensed by the Publisher, whether the whole or part of the material is concerned, specifically the rights of translation, reprinting, reuse of illustrations, recitation, broadcasting, reproduction on microfilms or in any other physical way, and transmission or information storage and retrieval, electronic adaptation, computer software, or by similar or dissimilar methodology now known or hereafter developed.
The use of general descriptive names, registered names, trademarks, service marks, etc. in this publication does not imply, even in the absence of a specific statement, that such names are exempt from the relevant protective laws and regulations and therefore free for general use.
The publisher, the authors and the editors are safe to assume that the advice and information in this book are believed to be true and accurate at the date of publication. Neither the publisher nor the authors or the editors give a warranty, expressed or implied, with respect to the material contained herein or for any errors or omissions that may have been made. The publisher remains neutral with regard to jurisdictional claims in published maps and institutional affiliations.

This Springer imprint is published by the registered company Springer Nature Switzerland AG
The registered company address is: Gewerbestrasse 11, 6330 Cham, Switzerland

If disposing of this product, please recycle the paper.

Preface

The present LNCS volume encloses research papers presented at the 29th International Database Engineered Applications Symposium (IDEAS), which was held on July 14–16, 2025, in Newcastle upon Tyne, UK. The symposium always provides an international forum for discussing the problems of engineering data-driven systems, involving not only database technology but also related areas such as artificial intelligence, communication, data mining, information retrieval, human-machine interaction, multimedia, natural language processing, privacy, security, and others. IDEAS always fosters closer interaction among the industrial, research, and user communities, providing an excellent opportunity for them to meet, discuss ideas, examine current ones, and develop new solutions and research directions. Along with the technical sessions, the symposium also regularly features prominent invited speakers.

This conference has been held annually since its first meeting in Montreal, Canada (1997). Other previous meetings were held at Bayonne, France (2024); Heraklion, Greece (2023); Budapest, Hungary (2022); online/Montreal, Canada (2021); online/Seoul, South Korea (2020); Athens, Greece (2019); Villa San Giovanni, Italy (2018); Bristol, UK (2017); Montreal, Canada (2016); Yokohama, Japan (2015); Porto, Portugal (2014); Barcelona, Spain (2013); Prague, Czech Republic (2012); Lisbon, Portugal (2011); Montreal, Canada (2010); Calabria, Italy (2009); Coimbra, Portugal (2008); Banff, Canada (2007); Delhi, India (2006); Montreal, Canada (2005); Coimbra, Portugal (2004); Hong Kong, China (2003); Edmonton, Canada (2002); Grenoble, France (2001); Yokohama, Japan (2000); Montreal, Canada (1999); and Cardiff, UK (1998).

The IDEAS 2025 conference received 30 original submissions from 11 different countries and four continents. The final set of selected papers included 13 long papers (43%) and 6 short papers (20%), a total of 19 papers. The selected articles span a wide spectrum of topics in the broader field of data management and were organized into five sessions: (1) Language and Models, (2) Classification, (3) Distributed Systems, (4) Query Answering and Education, and (5) Data Mining.

In addition, the IDEAS 2025 conference included two keynote talks that provided a broad-ranging view of topics in computer science and are as follows:

- Jim Webber (Neo4j), in his talk *"The Pub-Time Parliament"*, elaborated on the importance and role of consensus algorithms in building reliable systems;
- Laura Heels (Newcastle University) who dealt with the topic of *"Bias In, Bias Out: The Case for Protected Characteristics in Data"*.

The program also included two tutorials, which were delivered by:

- Mario Nascimento (Khory College's Director of Pacific Northwest Research, Northeastern University), who presented a tutorial entitled *"Approximate Nearest Neighbor Queries and Vector Databases"*;
- Hugo Firth and Jim Webber (Neo4j), who presented a tutorial entitled *"Modern Graph Databases: Practice and Research"*.

We are also grateful to the external sponsors who provided significant assistance in making the conference's social events possible: QTS Data Centres, Neo4j, Red Hat, and the National Innovation Centre for Data (NICD). These enabled a rich social program that included the following:

- A welcome reception at El Coto on the first day of the conference.
- A gala dinner at Blackfriars Restaurant on the following day.

The conference closed with an Industry Panel session on *"the role of databases in the age of AI"* which was led by Paul Watson (Newcastle University) with contributions from Louise Braithwaite (NICD), Mac Misiura (Red Hat), Anna Ollerenshaw (NVIDIA) and Jim Webber (Neo4j).

As customary from the former editions, the selected best papers of IDEAS 2025 will be invited for special issues of *Information* (MDPI). Therefore, the PC chairs express their gratitude to the Editors-in-Chiefs of the above journal for their approval regarding these special issues.

July 2025

Giacomo Bergami
Paul Ezhilchelvan
Yannis Manolopoulos
Sergio Ilarri
Jorge Bernardino
Carson K. Leung
Peter Z. Revesz

Organization

General Co-chairs

Jorge Bernardino   Institute of Education and Sciences (ISEC), Portugal
Carson K. Leung   University of Manitoba, Canada
Peter Z. Revesz   University of Nebraska-Lincoln, USA

Program Committee Co-chairs

Giacomo Bergami   Newcastle University, UK
Paul Ezhilchelvan   Newcastle University, UK

Tutorial Chair

Yannis Manolopoulos   Aristotle University of Thessaloniki, Greece

Organization Chair

Sergio Ilarri   University of Zaragoza, Spain

Steering Committee

Jorge Bernardino (Co-chair)   Institute of Education and Sciences (ISEC), Portugal
Rui Chen   Samsung, USA
Bipin Desai (Honorary Chair)   Concordia University, Canada
Le Gruenwald   University of Oklahoma, USA
Irena Holubova   Charles University, Czech Republic
Hideyuki Kawashima   Kyoto University, Japan
Bart Kuijpers   Hasselt University, Belgium
Carson K. Leung (Co-chair)   University of Manitoba, Canada
Yannis Manolopoulos   Aristotle University of Thessaloniki, Greece
Kamran Munir   University of the West of England, UK

Peter Z. Revesz (Co-chair) — University of Nebraska-Lincoln, USA
Jeffrey Ullman — Stanford University, USA

Program Committee

Alberto Freitas	University of Porto, Portugal
Alicja Wieczorkowska	Polish-Japanese Academy of Information Tech., Poland
Antonio Corral	University of Almería, Spain
Carson K. Leung	University of Manitoba, Canada
Chuan-Ming Liu	National Taipei University of Technology, Taiwan
Danilo Montesi	University of Bologna, Italy
David Chiu	University of Puget Sound, USA
Francesco Buccafurri	Università Mediterranea di Reggio Calabria, Italy
Giacomo Bergami	Newcastle University, UK
Giuseppe Polese	University of Salerno, Italy
Irena Holubova	Charles University in Prague, Czech Republic
Jaroslav Pokorny	Charles University in Prague, Czech Republic
Jeffrey Ullman	Stanford University, USA
Jorge Bernardino	Institute of Education and Sciences (ISEC), Portugal
Jose R. R. Viqueira	University of Santiago de Compostela, Spain
Marcos Aurelio Domingues	State University of Maringá, Brazil
Marinette Savonnet	University of Burgundy, France
Masayoshi Aritsugi	Kumamoto University, Japan
Miguel Rodriguez Luaces	Universidade da Coruña, Spain
Minal Bhise	DAIICT, Gandhinagar, India
Paulo Jorge Oliveira	ISEP, Portugal
Pavel Koupil	Charles University in Prague, Czech Republic
Richard Chbeir	University of Pau & the Adour Region, France
Roberto Yus	University of Maryland, Baltimore County, USA
Sabri Allani	Expleo Group, France
Sergio Ilarri	University of Zaragoza, Spain
Sven Groppe	University of Lübeck, Germany
Toshiyuki Amagasa	University of Tsukuba, Japan
Valeria Magalhaes Pequeno	INESC-ID, Portugal
Wilfred Ng	Hong Kong University of Science and Technology, China
Yannis Manolopoulos	Aristotle University of Thessaloniki, Greece
Yiu-Kai Ng	Brigham Young University, USA

External Reviewers

Adam Pazdor	University of Manitoba, Canada
Baha Rababah	Red River College Polytechnic & University of Manitoba, Canada
Connor Hryhoruk	University of Manitoba, Canada
Hoang Hai Nguyen	Canadian Food Inspection Agency & University of Manitoba, Canada
Oliver Robert Fox	Newcastle University, UK
Samuel Appleby	Newcastle University, UK

LogDS Lab Support Team

Oliver Robert Fox	Newcastle University, UK
Rohin Callum Percy Gillgallon	Newcastle University, UK

Contents

Language and Models

Generative Adversarial Networks Reveal Carian, Elder Futhark, Old
Hungarian and Old Turkic Script Relationships 3
    Shohaib Shaffiey and Peter Z. Revesz

EHSAN: Leveraging ChatGPT in a Hybrid Framework for Arabic
Aspect-Based Sentiment Analysis in Healthcare 17
    Eman Alamoudi and Ellis Solaiman

Automated Glyph Feature Detection Using Convolutional Neural Networks 34
    Michael Mason, Sam Kirchner, and Carter Powell

A Vision for Robust and Human-Centric LLM-Based QR Code Security 48
    Hissah Almousa and Ellis Solaiman

Classification

Exploring Classification with Spectral Transformation 61
    Alexander Stahl

Optimizing Classification Accuracy with Simulated Annealing
in k-Anonymity 71
    Despina Tawadros, Wenhui Yang, Lena Wiese, and Volker Meyer

Predicting Gelation in Copolymers Using Deep Learning Through
a Comparative Study of ANN, CNN, and LSTM Models with SHAP
Explainability 85
    Selahattin Barış Çelebi, Ammar Aslan, and Mutlu Canpolat

A Total Variation Regularized Framework for Epilepsy-Related MRI
Image Segmentation 96
    Mehdi Rabiee, Sergio Greco, Reza Shahbazian, and Irina Trubitsyna

Enhancing Flight Delay Prediction with Network-Aware Ensemble
Learning 109
    Mary Dufie Afrane, Yao Xu, and Lixin Li

Distributed Systems

FedMod: Vertical Federated Learning Using Multi-server Secret Sharing 125
  Kasra Mojallal, Ali Abbasi Tadi, and Dima Alhadidi

Throughput-Driven Database Replication Using a Ring-Based Order
Protocol 140
  Ye Liu, Paul Ezhilchelvan, Yingming Wang, and Jim Webber

Blockchain-Backed Fuzzy Search for Semi-structured Translation Data:
A Scalable Hybrid Approach with Hyperledger Fabric and Elasticsearch 155
  Edvan Soares and Valeria Times

Query Answering and Education

Towards Sustainable DBMS: A Framework for Real-Time Energy
Estimation and Query Categorization 171
  *Tidenek Fekadu Kore, David Sarramia, Myoung-Ah Kang,
  and François Pinet*

Context-Aware Visualization for Explainable AI Recommendations
in Social Media: A Vision for User-Aligned Explanations 184
  Banan Mohammad Alkhateeb and Ellis Solaiman

Transparent Adaptive Learning via Data-Centric Multimodal Explainable
AI 197
  Maryam Mosleh, Marie Devlin, and Ellis Solaiman

Analyzing Student Feedback to Assess NoSQL Education 211
  Vanessa Meyer, Lena Wiese, and Ahmed Al-Ghezi

Data Mining

Data Mining for Language Superfamilies Using Congruent Sound Groups 227
  Peter Z. Revesz and Mohanendra Siddha

A Cross-Linguistic Analysis of Linear A, Linear B and Swahili 242
  Joslin Ishimwe, Adrian Ratwatte, and Prince Ngiruwonsanga

Automated Identification of Allographs Among the Indus Valley Script
Signs 252
  Harsh Tamkiya, Gunjit Agrawal, Chiradeep Debnath, and Peter Z. Revesz

Author Index 263

Language and Models

Language and Model

Generative Adversarial Networks Reveal Carian, Elder Futhark, Old Hungarian and Old Turkic Script Relationships

Shohaib Shaffiey and Peter Z. Revesz

School of Computing, University of Nebraska-Lincoln, Lincoln, NE 68508, USA
{sshaffiey2,prevesz1}@unl.edu

Abstract. The precise classification and relationship of several ancient scripts has been the subject of debate for over a century. These controversial scripts include the Carian alphabet, the Elder Futhark alphabet, the Old Hungarian script, and the Yeniseian variant of the Old Turkic script. This paper settles the relationship among these controversial scripts in an objective and algorithmic way by using a Convolutional Neural Network augmented with a Generative Adversarial Network, which gives a probability of the membership of each sign in any script. The results yield a similarity metric between pairs of scripts, and that allows the mapping of the evolution of these scripts using a phylogenetic tree algorithm.

Keywords: Convolutional neural network · Generative adversarial network · Carian alphabet · Elder Futhark · Old Hungarian script · Old Turkic script · Yeniseian script

1 Introduction

The study of the evolution of various scripts is a fascinating but often controversial subject [11,13]. An original script and its descendants form a script family. The most well-known script family is the Phoenician Alphabet Family. Another script family that spread from the island of Crete is called the Cretan Script Family [21]. Both of these script families have a tendency of an increasing percentage of mirror symmetric signs [26] and often have other characteristics such as reflecting a front-back vowel harmony in the underlying language [23,30]. In contrast, some other scripts such as the Indus Valley Script [9] do not seem to have any known descendants. This paper focuses on the Carian alphabet, Elder Futhark, which is the oldest runic alphabet, the Old Hungarian script (in Hungarian: *székely-magyar rovásírás*), and the Yeniseian variant of the Old Turkic script, whose origins are still debated. For example, the Old Hungarian alphabet is hypothesized to derive from the Yeniseian script by Róna-Tas [27], a Turkic script by Sándor [28], from the Phoenician alphabet by Hosszú [16], and from the Carian alphabet by Revesz [22], who showed that some Hungarian speakers used the Carian alphabet between the 3rd and the 7th centuries [25]. On the other hand, tamga signs similar to Turkic tamga signs are sometimes found inserted into Old Hungarian inscriptions [24]. The origin of the Elder Futhark alphabet is also puzzling, with Hempl [15] advocating a Thracian alphabet origin following Isaac Taylor, while Mees advocating an Etruscan alphabet origin [19].

A large part of the controversy stems from manual studies of these controversial scripts, where the authors focus on a few signs that look similar to their eyes. In contrast, this paper gives an objective, algorithmic answer to such controversial questions. Daggumati and Revesz [10] initiated the study of the evolution of scripts by using convolutional neural networks (CNNs) [18]. They first trained separate CNNs to recognize various scripts [10]. Then they passed the signs of one script into a CNN that was trained to recognize another script. The sign that was recognized by the trained CNN would be the sign that closest matches the input sign. Two scripts can be considered related if there is a one-to-one mapping between their signs. Daggumati and Revesz [10] showed that the eight Bronze Age scripts that were used from Greece to India fall into three clusters. Scripts within each cluster are so closely related to each other that either they have a common origin, or one script is the origin of the other script in the cluster. For example, one of the interesting results of Daggumati and Revesz [10] is that Sumerian Pictograms, which preceded the later cuneiform script, are the likely origins of the Indus Valley Script, which is the oldest known writing in India and Pakistan.

The seminal work of Daggumati and Revesz [8,10] still left unresolved the relationship of numerous later scripts that first appeared in the Iron Age or later. There is a great need to continue their work to resolve the origin of these later scripts. From a methodological perspective, our novel contribution is that we will augment the dataset using Generative Adversarial Networks (GANs) [14] to generate examples of the signs of each script. This will lead to an augmented dataset on top of the on-the-fly transformations done with Python's libraries to mimic traditional methods to augment imaging datasets to use distortions (rotations, flips, etc.). The higher quality models that result from the GAN augmentation will be used in the training phase of the convolutional neural networks to classify the images more efficiently and accurately. We already achieve a high accuracy with only 10 epochs of training, while Daggumati and Revesz [10] had to go to 100 epochs of training to achieve a similar accuracy.

The rest of this paper is organized as follows. Section 2 describes the data sources, including a detailed description of each of the scripts considered, and reviews some basic definitions. Section 3 describes the experimental design and results. Section 4 discusses the implications of the experimental results regarding the evolution of the studied scripts. Finally, Sect. 5 presents some conclusions and directions for further work.

2 Data Sources and Basic Definitions

2.1 The Analyzed Scripts

The primary data sources of our work were the signs from the following scripts:

1. The Carian alphabet, which is shown in Fig. 1, was used between the 7th and 4th century BCE. It was used by the Carian people of Anatolia. It has been found in both Anatolia and Egypt, where Carians served as mercenaries [1].
2. The Elder Futhark alphabet, which is shown in Fig. 2, was used primarily between the 2nd and 8th centuries. It consists of 24 sign characters. The script engravings have been found on weapons, rune stones, and amulets from Scandinavia, Germany, the British Isles [7].
3. The Old Hungarian script, which is shown in Fig. 3, is a writing system that was used by the Hungarians before the adoption of the Latin Alphabet in the 11th century. It was used primarily among the Székely people in Transylvania [31].
4. The Yeniseian variant of the Old Turkic script, which is shown in Fig. 4, was used by Göktürks and other peoples of the Turkic khanates between the 8th and 10th centuries. The Yeniseian variant of the Old Turkic script developed from the earlier Orkhon variant, which appeared in the Orkshon Valley in Mongolia in the 8th century [12]. We chose the Yeniseian variant of the Old Turkic script, which developed from the Orkhon variant, because it was geographically closer to the usage area of the other scripts and therefore had more opportunities to influence them.

Fig. 1. The Carian alphabet.

Fig. 2. The Elder Futhark alphabet.

Unicode signs for the four scripts were extracted from Omniglot [2–5]. The images used for classification were unicode signs that were used as visual representations. Each character sign was augmented during training via rotation, flipping, etc. The signs were additionally used as inputs for GANs, which generated many high quality sign samples. A hundred samples of each sign were created by GANs to augment the dataset alongside the augmentation of each sign during training through character manipulations like translation, rotation, flipping, etc.

Fig. 3. The Old Hungarian script signs.

Fig. 4. The Yeniseian script signs.

2.2 Heatmaps, Staircases, and a Similarity Metric Between Scripts

The primary goal of our analysis is to generate a heatmap between any pair of scripts. The rows and the columns of a heatmap can be arranged in a staircase-like pattern if there is a one-to-one mapping between the signs of the two scripts. For example, Fig. 5 shows the heatmap that was generated when the Phoenician alphabet letters were passed into the CNN that learned the Greek alphabet [10]. The heatmap shows that when Phoenician *aleph* was passed into the Greek CNN, then it was recognized as the letter *alpha*. Similarly, the letters from Phoenician *beth* to *taw* were recognized as the Greek letters from *beta* to *tau*, respectively. The Greek alphabet has a few extra letters than denote vowels and diphthongs and those were not part of the emerging staircase-like pattern.

It is usually possible to start building a staircase for any pair of scripts. However, the staircase will have a maximal or near-maximal length only if the two scripts are related to each other. Figure 5 shows a maximal length staircase because it uses all of the twenty-two Phoenician letters. The existence of a (near)-maximal staircase may indicate an ancestor-descendant relationship like between the Phoenician alphabet and the Greek alphabet, where the former is well-known to be the ancestor of the latter. However, it may also indicate that they have a common ancestor.

If a staircase is generated when passing in script X to the CNN that learned script Y, then usually also a similar length staircase is generated in the reverse, that is, when passing in script Y to the CNN that learned X.

To develop a similarity metric *sim* between two scripts, we first fix a recognition strength threshold T, which decides which little square is part of the staircase. A square of the heatmap is part of the staircase if and only if its recognition value is above T. For example, all the little squares that form a staircase between the twenty-two Phoenician letters and their counterparts in the Greek alphabet would be counted if the threshold is $T = 0.5$. Let $s(a, b)$ be

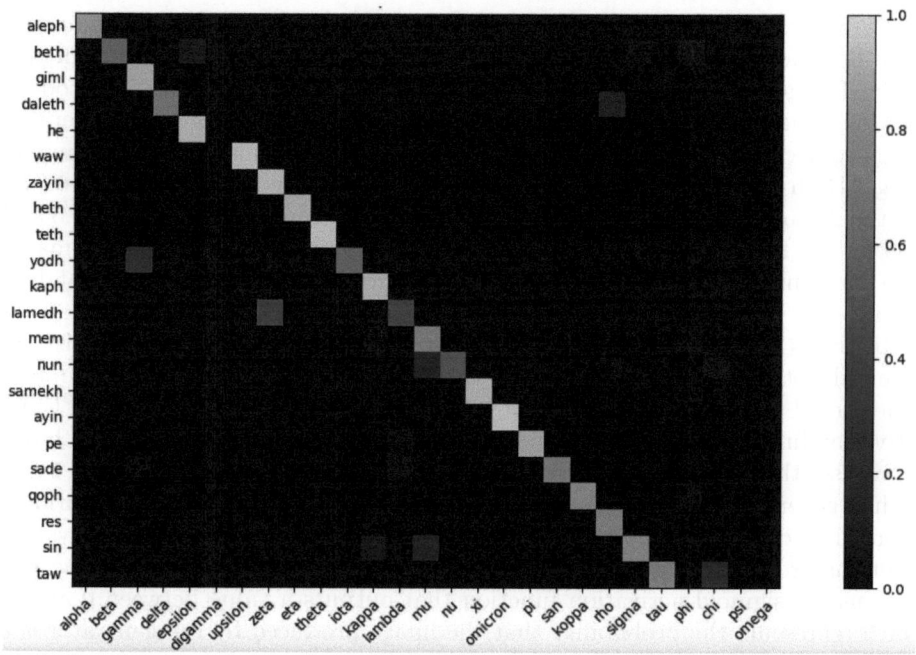

Fig. 5. The heatmap generated when passing in the Phoenician letters to the Greek CNN.

the strength of the recognition of sign a of script X as sign b of script Y. We only count mutually recognized signs. That is,

$$sim(X,Y) = \sum_{a \in X, b \in Y} (1 \ \ if \ \ s(a,b) \geq T \ \ and \ \ s(b,a) \geq T; \ \ \ 0 \ \ otherwise)$$

Since each pair of letters of the Phoenician and the Greek alphabets are mutually recognized [10], $sim(Phoenician, Greek) = 22$.

3 Experimental Design and Results

3.1 Experimental Design

Python, along with PyTorch libraries, was used to build separate CNNs for each script and GANs to augment the dataset. Each CNN architecture consists of a convolutional layer followed by max-pooling to downsample the image, then followed by two fully connected layers that map 256 hidden features to the number of signs in each script. A softmax is applied to the final output, representing the probability confidence of the predicted sign. The network uses a cross-entropy loss and uses the Adam optimizer [17] with a learning rate of 0.001. We used softmax instead of an SVM classifier because SVMs may have trouble with decision boundaries in larger datasets [6].

Python was used to build Generative Adversarial Networks (GANs) to generate images of individual script characters for each of the five scripts in this study. The generator takes in random noise as input and upscales it to produce 64 × 64 images. The discriminator uses the real script sign and the generated image to decide whether the generated image is indistinguishable from the real image. Both the discriminator and generator use binary cross-entropy loss and use the Adam optimizer with a learning rate of 0.0002 and beta values of 0.5 and 0.99 for the first and second moment terms, respectively. The generated images help augment the dataset and improve classification performance when real data is scarce.

In terms of the architecture, the generator of our GAN model consists of a neural network that contains four convolutional layers to upscale the 100-dimensional Gaussian noise to 64 × 64 images. ReLU (Rectified Linear Unit) activation functions are used to encourage gradient flow in all but the last layer, which uses the Tanh activation function to produce pixel values between -1 and 1 while generating images. The discriminator also consists of four convolutional layers that downscale the real and generated images. Leaky ReLU activation functions are used to allow gradients even for negative input values. The last layer uses a sigmoid activation function that outputs a value between 0 and 1, which represents the probability that the image produced by the generator is a real image.

3.2 Training Results

Figure 6 shows some examples of Yeniseian signs that were generated by our GAN after going through its full training cycle. The signs are only minor variations of the standard Yeniseian script sign.

Our GAN, which created one hundred images for each sign of each script for the training set together with our CNN architecture performed well in learning to recognize the signs of the five scripts. Table 1 shows some validation results. Each CNN achieved over 93% accuracy after ten iterations and over 98% accuracy after only twenty iterations. The accuracy kept slightly improving until the thirtieth iteration, when we stopped. Not surprisingly, the Old Hungarian script, which had the most number of signs, took slightly more iterations to converge than the CNNs associated with the other scripts to an over 99% accuracy.

3.3 Cross-Checking Experiment and Results

Next, we test the relationship among the pairs of scripts like in [10]. Hence, each script was passed through a trained CNN of a different script to assess the correlation of signs among the scripts. If we find a one-to-one mapping between two scripts, then they are related scripts.

Figure 7 shows the staircases that resulted when cross-checking the Carian alphabet with the Elder Furthark alphabet (top), the Old Hungarian script (middle), and the Yeniseian script (bottom). These staircases had eleven, ten and four steps, respectively.

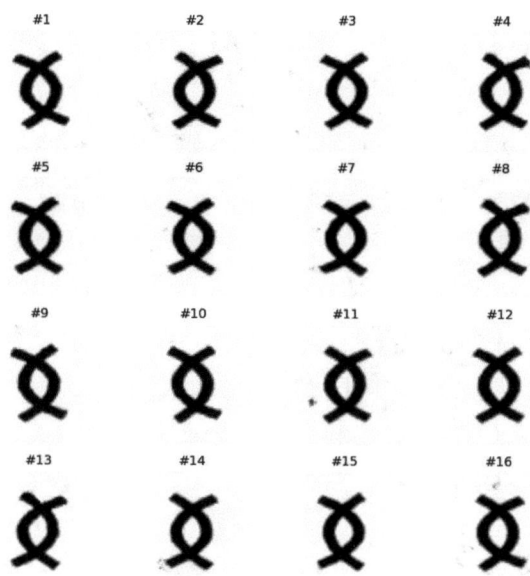

Fig. 6. GAN-generated examples of Yeniseian script signs.

Table 1. CNN recognition accuracy for the ancient script signs after training.

Script	Iteration 10	Iteration 20	Iteration 30
Carian	93.45	99.08	99.22
Elder Futhark	97.63	99.06	99.26
Old Hungarian	95.90	98.84	99.18
Yeniseian	96.72	99.01	99.30

Signs of different scripts are considered to be pairs if and only if each of their CNNs recognizes that the signs are the same at a threshold $T = 0.75$, that is, with at least 75% confidence. For example, the Carian letter denoting /n/ and the Elder Futhark letter denoting /z/ both have a trident-like shape. Hence, the CNN that learned the Carian alphabet recognizes the Elder Futhark letter /z/ as the Carian letter /n/, and the CNN that learned the Elder Futhark alphabet recognizes the Carian letter /n/ as the Elder Futhark letter /z/. Therefore, these two letters form a pair and are part of the staircases on the top of the figure. The staircases only show those signs that are mutually recognized.

Figure 8 shows three more examples of cross-checking pairs of scripts. The cross-checking between the Elder Futhark and Old Hungarian is shown on the top of the figure. This produced staircases with ten steps. The cross-checking between the Elder Futhark alphabet and the Yeniseian script is shown in the middle. This produced a simple staircases with only one step. Finally, the cross-

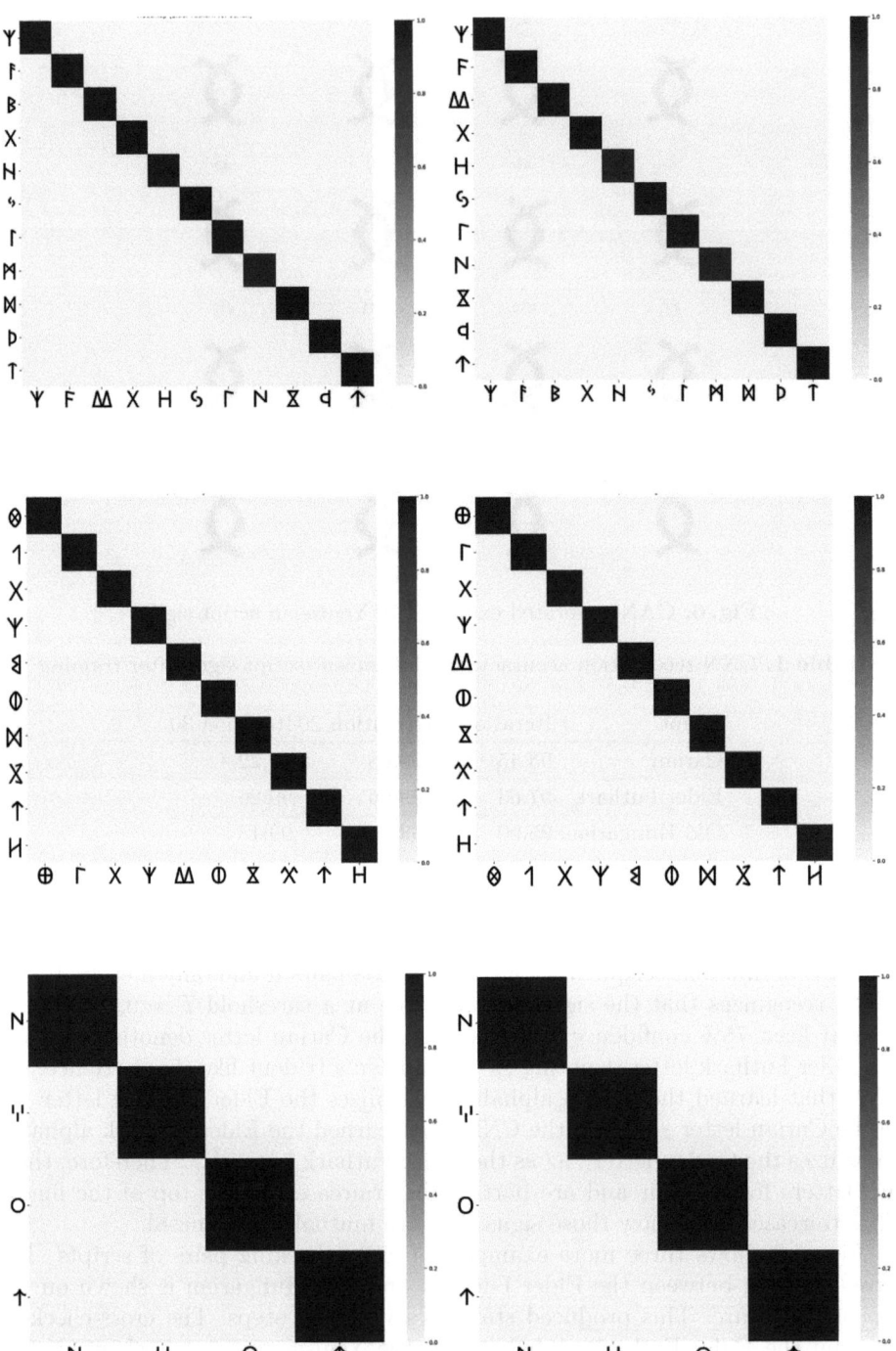

Fig. 7. Carian vs. Elder Futhark (top), Carian vs. Old Hungarian (middle), and Carian vs. Yeniseian (bottom). Those signs that did not mutually match are omitted.

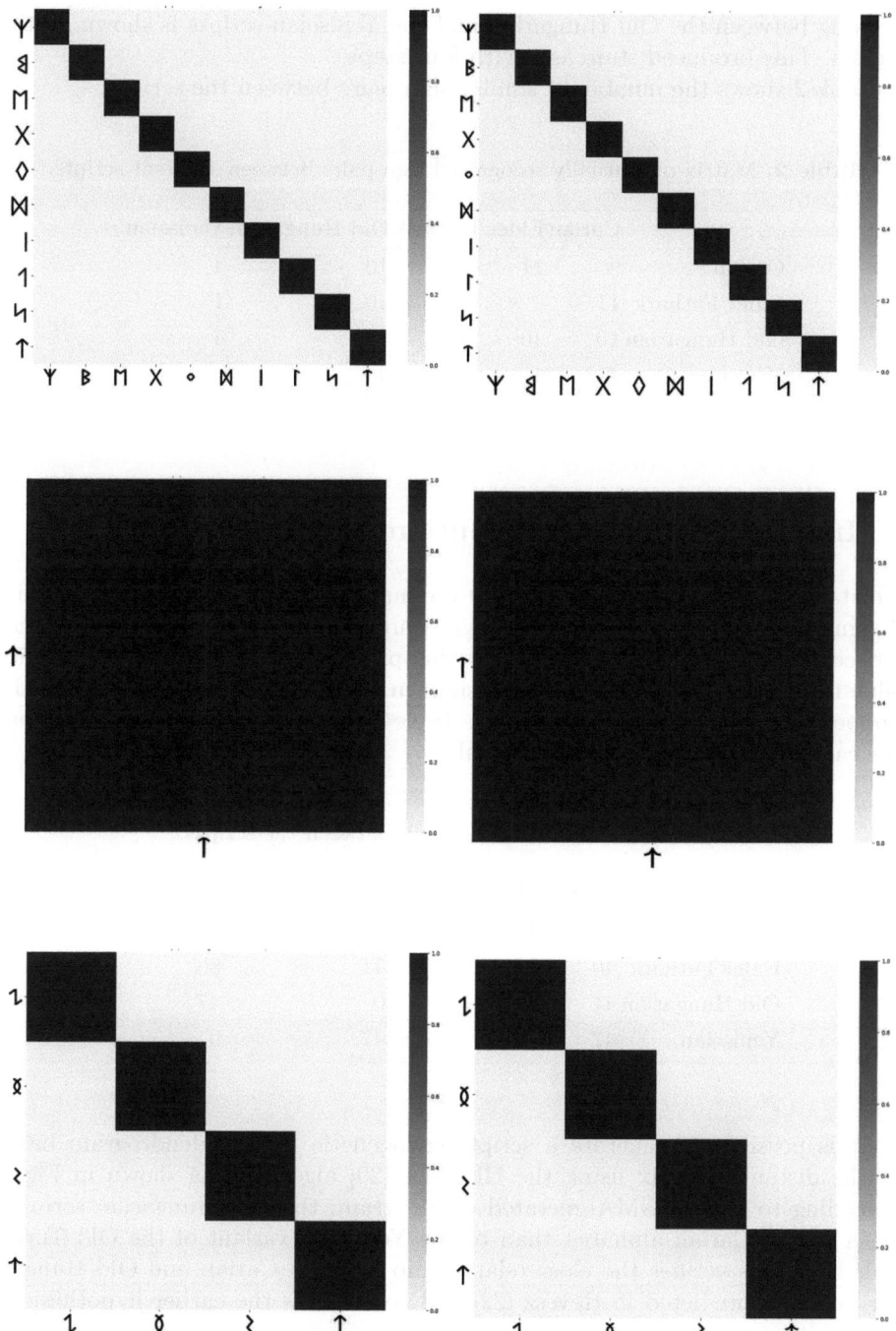

Fig. 8. Elder Futhark vs. Old Hungarian (top), Elder Futhark vs. Yeniseian (middle), and Old Hungarian vs. Yeniseian (bottom). Those signs that did not mutually match are omitted.

checking between the Old Hungarian and the Yeniseian scripts is shown on the bottom. This produced staircases with four steps.

Table 2 shows the number of similar sign pairs between the scripts.

Table 2. Matrix of mutually recognized sign pairs between different scripts.

	Carian	Elder Futhark	Old Hungarian	Yeniseian
Carian		11	10	4
Elder Futhark	11		10	1
Old Hungarian	10	10		4
Yeniseian	4	1	4	

4 Implications for the Evolution of the Scripts

A distance matrix can be derived from the number of mutually recognized pairs of signs shown in Table 2. The more signs are common the less is the distance between two scripts. In accordance to this principle, we can take the distance value to be the maximum number of signs in a script, which was 51, minus the number of mutually recognized signs. Table 3 shows the distance matrix that was calculated according to this formula.

Table 3. The distance matrix between the scripts.

	Carian	Elder Futhark	Old Hungarian	Yeniseian
Carian	0	40	41	47
Elder Futhark	40	0	41	50
Old Hungarian	41	41	0	47
Yeniseian	47	50	47	0

It is possible to generate a script phylogenetic tree or dendrogram based on the distance matrix using the UPGMA [29] algorithm as shown in Fig. 9. According to the UPGMA-gnerated dendrogram, the Old Hungarian script is closer to the Carian alphabet than to the Yeniseian variant of the Old Turkic alphabet. This verifies the close relationship between Carian and Old Hungarian that was predicted in Revesz [22] and contradicts the earlier hypothesis of

Róna-Tas [27] that the Yeniseian script is the ancestor of the Old Hungarian script. In fact, the Yeniseian script was the furthest from the rest of the scripts.

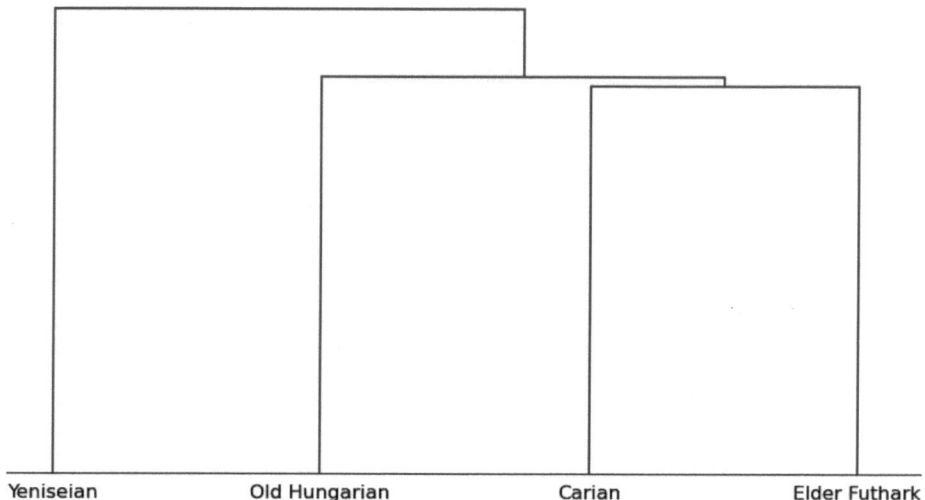

Fig. 9. The dendrogram generated by the UPGMA algorithm.

The UPGMA-generated dendrogram does not tell us what happened during the evolution of these scripts. Hence, we also applied the Common Mutation Matrix (CMM) algorithm of Revesz [20] to obtain a deeper understanding of the evolution of these scripts. The CMM algorithm takes as input the actual set of sign pairs and generates another dendrogram where on the edges we can see the shared signs.

The CMM-generated dendrogram is shown in Fig. 10. While the UPGMA and the CMM-generated dendrograms are structurally the same, it is possible to see when certain signs changed or were added to the alphabet. Similarly to the UPGMA algorithm, the CMM algorithm also assumes that the five scripts had a common ancestor, which is denoted by μ at the top of the dendrogram. However, the common origin of these scripts is supported only by a single shared sign, which is the up-arrow sign. Since the up-arrow sign is fairly common in scripts, it may be an accidental similarity.

The Carian alphabet, the Elder Futhark and the Old Hungarian alphabets share six additional signs that are missing from the Yeniseian script. That agrees with the hypothesis of Revesz [22] that Old Hungarian is closer to Carian than to the Yeniseian Old Turkic script. The Carian and the Elder Futhark share four additional signs that are missing from the Old Hungarian alphabet.

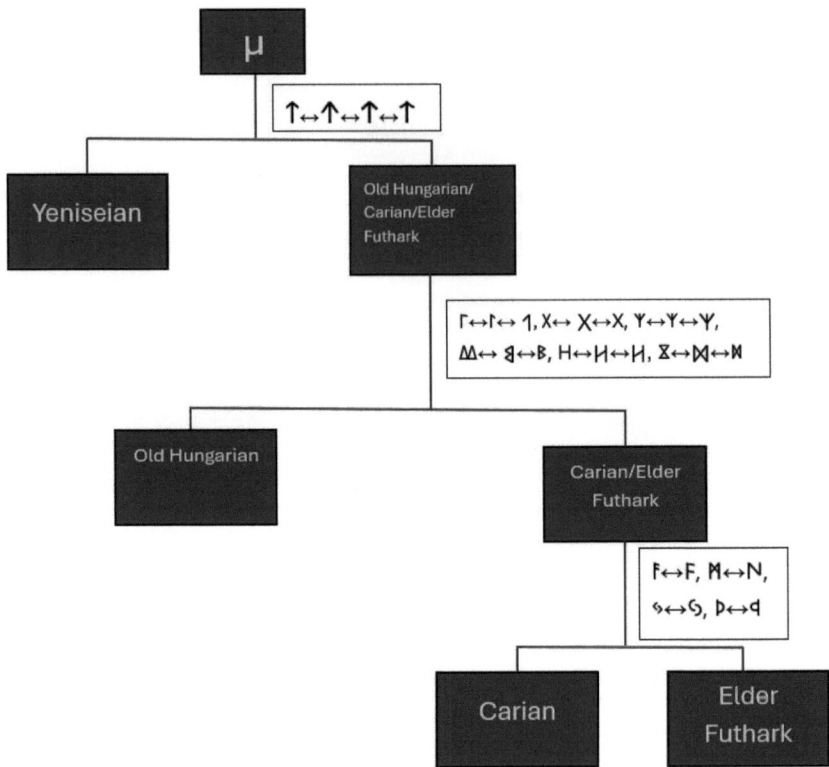

Fig. 10. The dendrogram generated by the CMM algorithm.

5 Conclusions and Future Work

This paper demonstrated that the inclusion of GAN-generated samples to augment the dataset markedly improved the accuracy and the convergence time for learning over using only real images and on-the-fly augmentations generated by manipulating the images by flipping, rotation, scaling, etc. Although the increased accuracy and convergence time are welcome, the potential problem of over-training by using GANs needs to be avoided. Hence, as a verification, we plan to use also the Variational AutoEncoders (VAEs) of Kingma et al. [17] to generate synthetic examples to augment the dataset and then compare the results with the current results.

While the CMM algorithm-generated dendrogram gave some details about the evolution of the scripts, it would be important to add additional scripts to gain a broader understanding of script evolution. Revesz [22] considered the Carian alphabet to be a descendant of the Linear A script, which was used by the Minoans on the island of Crete in the Bronze Age. We plan to test this hypothesis by adding the Linear A script to the set of scripts studied.

In addition, the evolution of the scripts needs to be examined from the perspective of phonetics too. That requires an interdisciplinary work together with phonologists because the phoneme denoted by some letters may have also changed as languages changed. It would give a deeper understanding of script evolution to show that most of the signs and their supposed predecessors are connected by explainable sound changes.

Disclosure of Interests. The authors have no competing interests to declare that are relevant to the content of this article.

References

1. Adiego, I.J.: The Carian Language, Handbook of Oriental Studies, vol. 73. Brill, Leiden (2007)
2. Ager, S.: Carian Alphabet (2024). www.omniglot.com/writing/carian.htm. Accessed 15 Mar 2025
3. Ager, S.: Elder Futhark (2024). www.omniglot.com/writing/runic.htm. Accessed 15 Mar 2025
4. Ager, S.: Old Hungarian (Székely-magyar rovás) (2024). www.omniglot.com/writing/oldhungarian.htm. Accessed 15 Mar 2025
5. Ager, S.: Old Turkic (Göktürk script) (2024). www.omniglot.com/writing/oldturkic.htm. Accessed 15 Mar 2025
6. Akram-Ali-Hammouri, Z., Fernández-Delgado, M., Cernadas, E., Barro, S.: Fast support vector classification for large-scale problems. IEEE Trans. Pattern Anal. Mach. Intell. **44**(10), 6184–6195 (2022). https://doi.org/10.1109/TPAMI.2021.3085969
7. Barnes, M.P.: Runes: A Handbook. Boydell Press (2012)
8. Daggumati, S., Revesz, P.Z.: Data mining ancient scripts to investigate their relationships and origins. In: Proceedings of the 23rd International Database Applications & Engineering Symposium, pp. 1–10. ACM Press
9. Daggumati, S., Revesz, P.Z.: A method of identifying allographs in undeciphered scripts and its application to the Indus valley script. Hum. Soc. Sci. Commun. **8**(50) (2021)
10. Daggumati, S., Revesz, P.Z.: Convolutional neural networks analysis reveals three possible sources of Bronze Age writings between Greece and India. Information **14**(4) (2023). https://doi.org/10.3390/info14040227
11. Daniels, P.T., Bright, W.: The World's Writing Systems. Oxford University Press (1996)
12. Erdal, M.: A Grammar of Old Turkic. Brill, Leiden (2004)
13. Fischer, S.R., Fischer, S.R.: History of Writing. Reaktion books (2003)
14. Goodfellow, I., et al.: Generative adversarial nets. In: Advances in Neural Information Processing Systems, pp. 2672–2680 (2014)
15. Hempl, G.: The origin of the runes. J. Ger. Philol. **2**(3), 370–374 (1899)
16. Hosszú, G.: Heritage of scribes: the relation of Rovas scripts to Eurasian writing systems. Rovas Foundation (2012)
17. Kingma, D.P., Welling, M.: Auto-encoding variational Bayes. arXiv preprint (2014). https://arxiv.org/abs/1312.6114

18. Krizhevsky, A., Sutskever, I., Hinton, G.E.: Imagenet classification with deep convolutional neural networks. In: Advances in Neural Information Processing Systems, pp. 1097–1105 (2012)
19. Mees, B.: The north Etruscan thesis of the origin of the runes. Ark. Nord. Filol. **115**, 33–82 (2000)
20. Revesz, P.Z.: An algorithm for constructing hypothetical evolutionary trees using common mutation similarity matrices. In: Proceedings of the 4th ACM Conference on Bioinformatics, Computational Biology and Biomedical Informatics, pp. 731–734. ACM Press (2013). https://doi.org/10.1145/2506583.2512391
21. Revesz, P.Z.: Bioinformatics evolutionary tree algorithms reveal the history of the Cretan Script Family. Int. J. Appl. Math. Inf. **10**(1), 67–76 (2016)
22. Revesz, P.Z.: Establishing the West-Ugric language family with Minoan, Hattic and Hungarian by a decipherment of Linear A. WSEAS Trans. Inf. Sci. Appl. **14**, 306–335 (2017). https://www.wseas.org/multimedia/journals/information/2017/a605909-068.pdf
23. Revesz, P.Z.: A vowel harmony testing algorithm to aid in ancient script decipherment. In: Proceedings of the 24th International Conference on Circuits, Systems, Communications and Computers (CSCC), pp. 35–38. IEEE Press (2020)
24. Revesz, P.Z.: Decipherment challenges due to Tamga and letter mix-ups in an Old Hungarian runic inscription from the Altai Mountains. Information **13**(9), 422 (2022). https://doi.org/10.3390/info13090422
25. Revesz, P.Z.: A tale of two sphinxes: Proof that the Potaissa Sphinx is authentic and other Aegean influences on early Hungarian inscriptions. Mediter. Archaeol. Archaeom. **24**(2), 191–216 (2024)
26. Revesz, P.Z.: Trend of increasing percentages of mirror-symmetric signs in the cretan script family and the Phoenician alphabet family. In: Understanding Information and its Role as a Tool: In Memory of Mark Burgin, pp. 445–465. World Scientific (2025)
27. Róna-Tas, A.: Hungarians and Europe in the Early Middle Ages: An Introduction to Early Hungarian History. Central European University Press, Budapest (1999). https://www.ceeol.com/search/book-detail?id=202183
28. Sándor, K.: A Turkic script in the Carpathian basin: the identity-marking functions of the Székely script. Acta Linguistica Academica **71**(4), 642–670 (2024)
29. Sokal, R.R., Michener, C.D.: A statistical method for evaluating systematic relationships. University of Kansas Scientific Bulletin **38**, 1409–1438 (1958)
30. VanOrsdale, J., Chauhan, J., Potlapally, S.V., Chanamolu, S., Kasara, S.P.R., Revesz, P.Z.: Measuring vowel harmony within Hungarian, the Indus Valley Script language, Spanish and Turkish using ERGM. In: Proceedings of the 26th International Database Engineered Applications Symposium (IDEAS), pp. 171–174. IEEE Press (2022)
31. Vásáry, I.: The Old Hungarian Script: History, Characteristics, and Revival. Hungarian Academy of Sciences Press, Budapest (2021)

EHSAN: Leveraging ChatGPT in a Hybrid Framework for Arabic Aspect-Based Sentiment Analysis in Healthcare

Eman Alamoudi[1,2](✉) and Ellis Solaiman[1]

[1] School of Computing, Newcastle University, Newcastle, UK
{E.S.O.Alamoudi2,Ellis.Solaiman}@newcastle.ac.uk
[2] Department of Information Technology, College of Computers and Information Technology, Taif University, Taif 21944, Saudi Arabia

Abstract. Arabic-language patient feedback remains under-analysed because dialect diversity and scarce aspect-level sentiment labels hinder automated assessment. To address this gap, we introduce EHSAN, a data-centric hybrid pipeline that merges ChatGPT pseudo-labelling with targeted human review to build the first explainable Arabic aspect-based sentiment dataset for healthcare. Each sentence is annotated with an aspect and sentiment label (positive, negative, or neutral), forming a pioneering Arabic dataset aligned with healthcare themes, with ChatGPT-generated rationales provided for each label to enhance transparency. To evaluate the impact of annotation quality on model performance, we created three versions of the training data: a fully supervised set with all labels reviewed by humans, a semi-supervised set with 50% human review, and an unsupervised set with only machine-generated labels. We fine-tuned two transformer models on these datasets for both aspect and sentiment classification. Experimental results show that our Arabic-specific model achieved high accuracy even with minimal human supervision, reflecting only a minor performance drop when using ChatGPT-only labels. Reducing the number of aspect classes notably improved classification metrics across the board. These findings demonstrate an effective, scalable approach to Arabic aspect-based sentiment analysis (SA) in healthcare, combining large language model annotation with human expertise to produce a robust and explainable dataset. Future directions include generalisation across hospitals, prompt refinement, and interpretable data-driven modelling.

Keywords: ChatGPT · Data Annotation · Aspect-Based Sentiment Analysis · Healthcare · Patient Satisfaction

1 Introduction

Patient feedback is increasingly recognised as a vital metric for evaluating and improving healthcare quality. Unlike standardised surveys, which often limit expression, free-form reviews, such as online comments and social media posts, allow patients to share more detailed and emotionally nuanced accounts of their experiences [1]. These narratives have proven valuable for identifying service strengths and weaknesses, complementing structured satisfaction tools, and informing healthcare [2–4]. The healthcare industry itself acknowledges this shift. According to a recent estimate, 90% of healthcare leaders view patient experience as a top strategic priority, while 45% identify improving satisfaction scores as a key goal – often through digital technologies that support more responsive, patient-centred care [5]. This growing interest has accelerated the adoption of natural language processing (NLP) tools, enabling large-scale analysis of narrative feedback and helping extract actionable insights for quality improvement. Among these tools, sentiment analysis (SA) has emerged as a central technique, categorising text based on emotional polarity – typically positive, negative, or neutral. In healthcare, SA supports the automatic classification of patient opinions, helping to detect patterns of satisfaction and dissatisfaction [2]. However, the same study found that SA in healthcare is less developed than in domains such as retail, primarily due to the complexity of medical narratives and the scarcity of annotated datasets. To address these limitations, many health systems rely on structured instruments such as HCAHPS in the US [6], and the Saudi Patient Experience Measurement Program in Saudi Arabia [7]. These tools, while useful, often constrain patient responses. For instance, patients may over-report their satisfaction due to gratitude bias, which can result in an overestimation of the actual quality of care [8]. In contrast, unsolicited narrative feedback can offer richer insights into care experiences, driving researchers toward more open-ended, patient-driven sources. In Arabic-language contexts, however, progress remains limited due to dialectal diversity, morphological complexity, and a scarcity of annotated data. Overcoming these challenges is crucial for developing inclusive NLP tools that can support quality healthcare analysis across diverse linguistic settings.

Our study addresses these gaps by introducing a fine-grained, sentence-level Arabic healthcare review analysis. In summary, the key contributions of this study are as follows:

- **Fine-Grained Sentence-Level Aspect-Based Sentiment Analysis (ABSA).** Arabic hospital reviews are segmented into individual sentences, each annotated with a specific service aspect and its corresponding sentiment. This enables SA at multiple levels, including both the document level and the more detailed aspect level within the text.
- **Hybrid Annotation Framework.** We introduce a three-tier annotation strategy combining ChatGPT pseudo-labelling with human review, producing fully supervised, semi-supervised, and unsupervised training sets. This design enables us to evaluate the impact of human oversight on model performance.
- **Explainable Annotations.** Each automated annotation is accompanied by a rationale generated by ChatGPT, improving transparency and allowing verifiability of the model's decisions.
- **Dual Taxonomy Levels.** We define two aspect category schemes – an initial 17-category taxonomy and a consolidated 6-category schema – to examine the effect of granularity on performance and to provide both detailed and high-level analysis options.
- **Empirical Evaluation with Transformer Models.** We implemented our explainable healthcare sentiment annotation (EHSAN) dataset and report comprehensive experiments fine-tuning an Arabic-specific BERT model (AraBERT) and a lightweight

multilingual model (DistilBERT) on our data under different supervision levels and classification granularities. This provides insights into how a domain-specific model compares to a general-purpose model on Arabic ABSA, as well as how much manual annotation is needed for high accuracy.
- **Validation of Pseudo-Labelling in Low-Resource NLP.** Our results empirically demonstrate that high-quality pseudo-labels from a Large Language Model (LLM) can serve as a reliable alternative to manual labels in a low-resource setting, with minimal loss of accuracy. This highlights a cost-effective path for building language-specific resources when expert annotation is scarce.
- **Data-Management Perspective.** We articulate how schema evolution, provenance capture and privacy-aware indexing turn ABSA into a data-engineering problem of direct interest to the IDEAS (International Conference on Intelligent Data Engineering and Automated Systems) community.

The remainder of this paper is organised as follows. The literature review examines existing research on SA in the healthcare sector. The methodology section describes the processes used for data collection, preprocessing, and topic modelling of Arabic healthcare reviews, as well as the details of the annotation workflow using ChatGPT, including prompt design and tiered annotation, and the details of the training models used, along with the evaluation metrics applied to assess their performance. The results section presents the main findings, which are further interpreted in the discussion section. Finally, the conclusion and future work section summarises key insights and proposes directions for future research.

2 Literature Review

SA in Healthcare. Sentiment analysis (SA) has been applied in healthcare to automatically gauge patient satisfaction from text. However, as [2] found, healthcare SA lags behind other domains due to limited training data and the complexity of medical narratives. Traditional patient experience measures, including the HCAHPS survey in the United States [6], and the Saudi Patient Experience Measurement Program [7], typically quantify satisfaction but provide limited qualitative insight. In contrast, unsolicited feedback (e.g., online reviews, social media) can capture rich, emotionally charged experiences. For example, [9] analysed thousands of Hungarian health forum posts to identify common complaints (waiting times, communication issues, etc.). Such studies underscore the value of narrative data for uncovering issues not evident in surveys. However, challenges in processing this data include casual language, misspellings, and mixed sentiments.

Aspect-Based Sentiment Analysis (ABSA). ABSA extends SA by linking sentiments to specific aspects or topics. In healthcare, this means identifying which part of the service a sentiment refers to (e.g., nursing staff, cleanliness, billing, etc.). [10] applied ABSA to 30,000 Indian hospital reviews, extracting aspect-specific sentiments (doctors, staff, facilities, cost) and showed that such granularity provides actionable insights for hospital management. In terms of the Arab world, research is nascent. [11] built a Saudi patient comments classifier (PX_BERT) that is able to categorise feedback into predefined aspect categories, but they did not perform sentiment labelling. [4] created datasets from Saudi Twitter and Google reviews using ChatGPT with human verification to annotate aspects such as medical staff, appointments, and the like, along with sentiment. Their HEAR dataset segmented feedback by aspect, but still operated on whole reviews (multi-label per review) rather than true sentence-level segmentation. Table 1 presents a summary of the Arabic datasets used in previous healthcare SA studies, along with the dataset used in the present study.

Table 1. Overview of Arabic datasets used in healthcare sentiment analysis studies.

Study	Dataset Name and Size	Granularity	Annotation Details	Annotation Rationale	Public Availability
[11]	PX dataset: labeled comments: 19,000, negative-only subset: 13,000	Comment-level (multi-label classification over 25 SHCT categories; no sentiment)	Manually labelled by Saudi Ministry of Health (MOH)	No	No – Internal MOH dataset
[4]	HoPE-SA: 12,400 tweets and HEAR: 25,156 reviews	Aspect-level (up to 5 predefined aspects per review; each assigned an individual sentiment)	ChatGPT + human verification	No	Yes
Present study	EHSAN, 6000 reviews	Sentence-level: Reviews segmented into sentences; each sentence labelled with one aspect (from 17 fine-grained or 6 coarse-grained categories) and sentiment. Aspect-level information is derived through aggregation across sentences.	ChatGPT + human verification	Yes	Yes

LLMs for Annotation. The advent of powerful LLMs, such as ChatGPT, has opened up new avenues for dataset creation in low-resource languages. Consequently, researchers have begun using LLMs to generate labels or augment data. For example, [12] applied few-shot prompting with ChatGPT to label emotion in French texts, achieving an F1 of 0.66 and outperforming classical classifiers. [13] used ChatGPT-4o to analyse Chinese patient reviews with high accuracy (F1 = 0.912), highlighting its ability to understand specialised and non-English texts. In Arabic healthcare NLP, [14] applied few-shot prompting with ChatGPT on a small Arabic SA corpus, reporting promising results and calling for a more comprehensive evaluation across tasks and datasets.

Despite these advances, challenges remain. While previous studies have introduced Arabic healthcare review datasets, none have provided the level of detail found in this work, particularly in terms of the granularity of aspect categories and the inclusion of explanations generated by large language models (LLMs). Furthermore, concerns about the content generated by LLMs, such as potential bias or inaccuracies, necessitate careful validation. This study addresses these gaps by developing the EHSAN dataset, which incorporates a rigorous annotation scheme and evaluates the extent to which manual correction was needed to train effective models.

3 Methodology

A multi-stage pipeline was developed using ChatGPT to pseudo-label sentences in Arabic hospital reviews with aspect categories and sentiment, supported by varying levels of human review. This approach enables the capture of issue-specific sentiments within individual reviews and ensures alignment with Saudi Arabia's healthcare complaint taxonomy. The outcome was the explainable healthcare sentiment annotation (EHSAN), a dataset that provides explainable, multi-level annotations for Arabic patient feedback.

3.1 Data Collection and Description

A broad dataset of patient reviews was collected from the Google Maps listings of hospitals in Saudi Arabia. The dataset comprised reviews of hospitals across various regions of the country, reflecting diverse regional dialects and vocabularies. Each hospital entry included a list of reviews containing text, star ratings, and metadata. Specifically, reviews were chosen from three representative hospitals – one each from the central region (the city of Riyadh), the western region (the city of Jeddah), and the eastern region (the city of Dammam) regions – to ensure geographic and dialectal diversity. The three hospitals also differed in their average patient ratings (ranging from approximately 2.4 to 4.0 out of 5), thereby capturing a range of both positive and negative patient experiences. All three hospitals selected were officially accredited and listed in the national registry valid through 2025 [15].

A total of 5,000 reviews were included at this stage, selected from the three target hospitals. Following sentence segmentation (a process described in a later section), the reviews yielded 9,337 individual sentences. From this pool, 2,000 sentences were randomly selected from each hospital, resulting in a total of 6,000 sentences used for detailed analysis.

All data were collected in adherence to ethical standards. Only publicly available review text was used, and any personal identifiers were either absent or removed to protect reviewer privacy. The final dataset (the explainable healthcare sentiment annotation, or EHSAN) was constructed to support aspect-based SA at the sentence level, with each sentence annotated with a specific service aspect and its corresponding sentiment. Additionally, the dataset supports document-level analysis by including an overall sentiment label for each complete review, allowing for flexible use across various analytical scopes.

The study protocol was reviewed and exempted by Newcastle University Ethics Committee (Ref: 59967/2023). It also fully complies with Google Maps Terms of Service.

3.2 Data Pre-processing

The raw review texts, as collected, were found to contain various inconsistencies and noise typical of user-generated content, particularly in Arabic. To address these issues, a two-phase pre-processing pipeline was implemented to clean and standardise the text prior to analysis, as explained below.

Structural Clean-Up. In the first phase, attention was given to text segmentation and formatting. Regular expressions were applied to insert explicit sentence breaks at punctuation marks (e.g., periods, question marks, and exclamation points), ensuring that each sentence appeared on a separate line. This step was essential due to the informal nature of many Arabic reviews, which often lack clear sentence boundaries. Emojis were removed and replaced with either placeholders or newlines to prevent interference with text processing. Excess whitespace was normalised by collapsing multiple newlines and spaces, thereby promoting formatting consistency. Any empty or null entries were discarded following this cleaning step.

Linguistic Normalisation. The second phase focused on addressing Arabic-specific textual variations. The AraBERT Pre-processor toolkit from AUBMindLab was employed to standardise the Arabic text. Key transformations included the removal of diacritics (Tashkeel), normalisation of variant ya letters (e.g., converting forms like "ي" to the standard "ى"), and unification of other character variants to ensure consistent spelling. Punctuation that did not serve a syntactic purpose was eliminated, and sequences of repeated letters were reduced to manageable lengths (e.g., "مرررررحبا" was shortened to "مررحبا") to address exaggeration patterns typical of informal writing. Similarly, the decorative elongation of letters was removed. These steps, informed by best practices in Arabic NLP, were applied to reduce noise and enable the model to focus on the core content.

Following pre-processing, the text was rendered significantly cleaner and more standardised while maintaining the original meaning of the reviews. This thorough process ensured that the subsequent segmentation and labelling stages could be applied consistently, despite variability in writing styles across the dataset.

3.3 Topic Modelling and Sentence Segmentation

One novel aspect of the methodology used in the present study is employing topic-driven sentence segmentation performed by ChatGPT (specifically, the GPT-4o-mini model). Instead of relying solely on punctuation-based splitting – which can be unreliable for informal text – ChatGPT was guided to segment each review such that each resulting sentence corresponded to a single topical idea or aspect. This was achieved by prompting ChatGPT with tailored instructions to divide the review into meaningful sentences, allowing for the merging or splitting of phrases as needed to isolate distinct topics. When a sentence in the original text addressed multiple topics, it was broken down into separate sentences, each focusing on a single topic. Conversely, when multiple short sentences pertained to the same topic, they were merged to enhance coherence. This approach ensured that each segmented sentence could later be assigned a single relevant aspect category with accuracy.

To inform and support this process, topic modelling was conducted on the corpus as an analytical tool – used not directly in the labelling pipeline but to validate topic coverage. BERTopic, an advanced topic modelling technique, was employed, which is capable of clustering semantically similar texts and extracting representative keywords for each cluster. When applied to the pre-processed reviews (prior to segmentation), BERTopic identified approximately 25 distinct topics covering themes such as staff attitude, waiting times, and facility cleanliness. These unsupervised insights were strongly aligned with the predefined taxonomy of 17 aspect categories, thereby providing validation for the comprehensiveness of the chosen taxonomy.

For comparative purposes, a rule-based Arabic NLP tool, Stanza [16], was used for sentence splitting. It was observed that ChatGPT's segmentation yielded higher topical coherence, while Stanza occasionally retained compound sentences or split text at every punctuation mark, regardless of semantic context. Based on these findings, the segmentation produced by ChatGPT was adopted for the final dataset. Each segmented sentence was then passed on to the annotation stage.

3.4 Annotation Strategy with ChatGPT (Pseudo-Labelling)

Following segmentation, each sentence – now treated as an individual review – was annotated by ChatGPT with an aspect category and sentiment label using a few-shot learning approach. The prompt provided to ChatGPT included detailed instructions along with definitions of the aspect categories and sentiment labels. For each review, two labels were requested: (1) the main topic or aspect of the review, selected from a predefined list of 17 categories (e.g., "Medical_and_Nursing_Staff," "Billing_and_Finance"), and (2) the sentiment expressed – positive, negative, or neutral.

Additionally, ChatGPT was instructed to provide a brief justification for each classification decision, explaining the rationale behind its choice of topic and sentiment based on the content of the sentence. This justification was initially generated in English because the model demonstrated higher reliability in producing accurate reasoning in English. Subsequently, a translation of the justification into Arabic was requested.

To enhance clarity and analytical precision, each sentence generated by ChatGPT was treated as an independent review. Nevertheless, the original review IDs were preserved and linked to all extracted sentences to ensure traceability and facilitate flexible future use. This structured design allows for analysis at multiple levels – sentence, aspect, or document – and ensures the adaptability of the dataset.

Human validation was applied across all dataset splits—namely, the training, validation, and test sets. The training data were fully reviewed and categorized into three supervision levels: fully verified, partially reviewed, and model-labelled without correction. Human annotators were native Arabic speakers, all with higher education degrees, understanding of the healthcare context, and familiarity with digital review environments. All annotators participated in a structured training programme prior to the annotation task, which included a workshop on the 17-aspect taxonomy and hands-on exercises using a pre-labelled subset of 50 sentences. To ensure annotation quality, they were required to pass a qualification task before contributing to the dataset. The level of inter-annotator agreement was high, with disagreements occurring in only three instances, all of which involved borderline or ambiguous cases. Annotation was performed by two primary annotators working collaboratively, with data entered into a structured Excel sheet designed to review and evaluate ChatGPT's auto-generated labels. In cases of disagreement, a third reviewer adjudicated the final label using a majority voting protocol. In addition, a subset of model justifications was reviewed and found clear and consistent with final labels, reflecting strong alignment with the model's rationale. To evaluate the reliability of this manual annotation protocol, a stratified random sample of 600 sentences, constituting 5% of the 12,000 total annotations (6,000 for topics and 6,000 for sentiments), was selected for inter-annotator agreement analysis. Prior to adjudication, each sentence was independently annotated by the two primary reviewers. Interrater agreement was quantified using Cohen's Kappa (κ) [17], which yielded a score of 0.873, thus reflecting almost perfect agreement. Nevertheless, human reviews were incorporated to varying degrees, as described in subsequent sections. Table 2 presents the accuracy of ChatGPT in classifying topics and sentiments across the training, validation, and test sets, with all results verified through human annotation.

Table 2. ChatGPT accuracy in topic and sentiment classification.

Dataset	Topic	Sentiment	Total Cases
Training	0.90	0.96	3600
Validation	0.93	0.96	1200
Testing	0.90	0.94	1200

3.5 Dataset Construction: Supervision Levels

A core component of the methodology involved evaluating varying levels of human involvement in the labelling process. To facilitate this, three versions of the training dataset were constructed:

Fully Supervised Dataset (FSD). All labels generated by ChatGPT (both aspect and sentiment) were fully reviewed and corrected by human annotators. This version represents a gold standard dataset in which every label was verified and considered accurate.

Semi-Supervised Dataset (SSD). In this version, half of the reviews in the training data were reviewed and corrected by human annotators, while the remaining half retained the original labels assigned by ChatGPT without modification. The human-reviewed portion was selected randomly to ensure that all aspect categories were represented and that no class bias was introduced.

Unsupervised Dataset (USD). No human corrections were applied in this version; the dataset relied entirely on ChatGPT's pseudo-labels. This variant was used to assess the model's performance when trained solely on machine-generated annotations.

The dataset was partitioned into three subsets: 3,600 reviews were allocated for training, 1,200 reviews were reserved for validation, and the remaining 1,200 were allocated for testing. The training subset was used as the foundation for generating the three versions described above, and each was subjected to a different level of human supervision. In contrast, all reviews within the validation and test subsets were fully annotated by human reviewers to ensure the presence of reliable ground truths for model evaluation. Due to the presence of English text in some reviews, any review in which English characters exceeded 25% of the total content was excluded. As a result, the final number of reviews in each subset became 3,583 for training, 1,189 for validation, and 1,190 for testing.

Additionally, two classification schemes were implemented for aspect categories within each dataset – one using the original 17 fine-grained classes, and another using six broader, consolidated classes. The 6-class scheme was developed by grouping semantically related fine-grained aspects, particularly where overlaps in meaning were observed. For instance, individual departments, such as radiology and surgery, were grouped under a unified "medical services" category. This grouping was introduced after it was observed that certain categories within the 17-class taxonomy were underrepresented (e.g., only four reviews pertained to privacy), and that even human annotators occasionally confused similar categories.

The six consolidated aspect categories were defined as medical services, administrative services, appointment and waiting, environment and facilities, billing and finance, and miscellaneous. This dual dataset structure – incorporating both 17-class and 6-class schemes – was designed to examine how classification granularity impacts model performance while also offering a more practical schema for applications in which high-level categorisation is sufficient.

3.6 Model Training and Evaluation

The annotated datasets were evaluated by fine-tuning two transformer-based models to perform multi-class topic classification and SA:

AraBERT. A BERT-language model pre-trained specifically on large Arabic corpora [18]. We used the AraBERT v0.2 (large) model, which has demonstrated strong performance on various Arabic NLP tasks and is well suited to handling both Modern Standard Arabic and dialectal Arabic text. We expected AraBERT to have an advantage in understanding the nuances of Arabic patient reviews.

DistilBERT (multilingual). A distilled (lightweight) version of BERT that supports multiple languages [19]. We chose the base multilingual DistilBERT to represent a more computationally efficient model. While it is not specialised in Arabic, it covers Arabic in its training and is much smaller than AraBERT. The comparison between AraBERT and DistilBERT allowed us to discern the trade-off between a large Arabic-specific model and a smaller general-purpose model for this task.

The classification models were fine-tuned using open-source frameworks, primarily Hugging Face and PyTorch, and all experiments were conducted on Paperspace using high-performance GPUs to accelerate training. We treated **aspect category classification** (17-class or 6-class) and **sentiment classification** (3-class) as two separate tasks. In practice, we trained one model for aspect classification and another for sentiment classification, rather than a single multi-task model. This was to simplify training and because the aspect taxonomy changed (17 to 6) whereas sentiment remained the same. Each model was trained on the training set (FSD, SSD, or USD) accordingly, and evaluated on the common fully supervised test data. We used the same data splits for both models to ensure comparability.

Training Details. Standard fine-tuning procedures were applied. A batch size of 4 and a learning rate of $1e-5$ were used, and fine-tuning was conducted for 5 epochs in each run; these hyperparameters were tuned on the validation set. Each model's own tokenizer was ensured: AraBERT's tokenizer was used for Arabic text, and DistilBERT's multilingual tokenizer was used for that model. Appropriate padding and truncation were applied to ensure that the input sequences conformed to the models' length requirements. Early stopping on the validation set was employed to prevent overfitting.

For evaluation, the macro-averaged F1 score was primarily reported, along with accuracy, macro-precision, and macro-recall for completeness. Macro-averaging (i.e., averaging metrics equally across all classes) was deemed appropriate due to class imbalance, ensuring that performance on rare classes was accounted for. Emphasis was placed on the F1 score, as it balances precision and recall – particularly important in a multi-class imbalanced scenario. Accuracy alone could be misleading if the model disproportionately favours majority classes. Training time (in minutes) was also measured for each model in each dataset to assess computational efficiency.

4 Results

After the models were trained, their performance was analysed on the held-out test set. The results for the aspect (topic) classification task under both the 17-class and the 6-class schemes are presented below, followed by the sentiment classification results. All results were reported based on the fully human-annotated test set to ensure fairness in the evaluation.

4.1 Aspect Classification Performance (17 Classes)

Table 3 summarises the performance of AraBERT and DistilBERT on the 17-category aspect classification task for each training dataset variant (FSD, SSD, and USD). Both models achieved reasonably good accuracy and F1, given the difficulty of the task (17 imbalanced classes). AraBERT outperformed DistilBERT in all scenarios. For instance, with fully supervised training data, AraBERT reached 79% accuracy and an F1 score of

0.66, compared to DistilBERT's 72% accuracy and 0.61 F1. The performance gap persisted in semi-supervised and unsupervised settings, indicating the benefit of a language-specific model for this task. Importantly, the difference between using fully human-reviewed data and relying solely on ChatGPT labels was insignificant, with AraBERT's F1 dropping from 0.66 (FSD) to 0.64 (USD), and DistilBERT's falling from 0.61 to 0.59. The semi-supervised case was intermediate. Minor improvements were seen with human corrections, especially for DistilBERT (which improved by about 0.02 in F1 with full supervision). Overall, the FSD yielded the best performance for both models, followed by the SSD, and the USD (machine-only) was only slightly behind. Notably, the AraBERT model trained on the USD (pure ChatGPT labels) even outperformed the DistilBERT model trained on the FSD (fully human labels).

To assess the reliability of the model results and mitigate the effect of random variation, a two-step statistical analysis was conducted. First, 95% confidence intervals were estimated via bootstrapping to evaluate the stability of F1 scores. Second, an Approximate Randomization Test was done to assess the statistical significance of performance differences. Based on five independent runs with varied random seeds, the results showed consistent confidence intervals in most cases, except for AraBERT on the USD dataset, which exhibited greater variability ([0.46, 0.65]). In contrast, its confidence interval on the FSD dataset was narrow ([0.66, 0.68]), indicating stable and reliable performance. Statistically significant differences were observed only between the FSD and USD settings in AraBERT ($p = 0.021$), while other comparisons showed no significant differences. Statistical testing was applied to the 17-aspect setup due to observable variability, unlike the 6-aspect and sentiment results, which showed near-identical scores.

4.2 Aspect Classification Performance (Six Classes)

After merging the aspect labels into six broader categories, the model performance improved markedly. Table 4 shows the results of the 6-class classification. With fewer classes and more training examples per class (since we combined several related classes), the models achieved higher scores across all metrics. AraBERT, in particular, reached 81% accuracy and 0.78 F1 on the FSD, a substantial increase over the 17-class scenario. DistilBERT also improved to approximately 75% accuracy, to 0.69 F1 in the FSD case. Strikingly, AraBERT's performance was identical for the FSD and the SSD in the 6-class setup – 0.78 F1 in both cases. Even in the USD scenario (no human labels), AraBERT maintained an F1 of around 0.77. DistilBERT showed a slight drop from the FSD (0.69 F1) to the SSD (0.67) to the USD (0.67). These results affirm that reducing the label noise (via human review) has diminishing returns once the classes are made easier (6 instead of 17) – the pseudo-labels were good enough that cleaning half or all of them gave minimal additional benefit, at least for the stronger model.

Table 3. Aspect classification results (17 classes).

Dataset	Model	Accuracy	Macro Precision	Macro Recall	Macro F1 Score	*Confidence Interval 95%*
FSD	AraBERT	0.79	0.65	0.69	0.66	[0.66, 0.68]
	DistilBERT	0.72	0.60	0.62	0.61	[0.61, 0.61]
SSD	AraBERT	0.80	0.65	0.69	0.65	[0.64, 0.67]
	DistilBERT	0.71	0.60	0.61	0.60	[0.60, 0.61]
USD	AraBERT	0.78	0.64	0.68	0.64	[0.46, 0.65]
	DistilBERT	0.71	0.59	0.61	0.59	[0.59, 0.61]

Table 4. Aspect classification results (6 classes).

Dataset	Model	Accuracy	Macro Precision	Macro Recall	Macro F1 Score
FSD	AraBERT	0.81	0.78	0.78	0.78
	DistilBERT	0.75	0.69	0.68	0.69
SSD	AraBERT	0.81	0.78	0.78	0.78
	DistilBERT	0.73	0.68	0.67	0.67
USD	AraBERT	0.81	0.79	0.77	0.77
	DistilBERT	0.73	0.69	0.67	0.67

4.3 Sentiment Classification Performance

In addition to aspect prediction, sentiment classification (positive, negative, or neutral) was also performed for each review within the framework. Separate sentiment classifiers were trained on the same training splits. Strong performance was demonstrated by both AraBERT and DistilBERT, with AraBERT consistently achieving superior results across all datasets. On the FSD split, AraBERT achieved an accuracy of 93%, along with a macro F1 score of 0.84, while DistilBERT yielded lower accuracy (81%) and an F1 score of 0.67. Comparable trends were observed with the SSD and the USD; AraBERT maintained a stable F1 score of 0.84, indicating robustness, even when trained on pseudo-labelled data. DistilBERT's performance remained consistent but lower, with F1 scores ranging from 0.66 to 0.67 across all splits. These findings are summarised in Table 5.

Table 5. Sentiment classification results.

Dataset	Model	Accuracy	Macro Precision	Macro Recall	Macro F1 Score
FSD	AraBERT	0.93	0.85	0.82	0.84
	DistilBERT	0.81	0.66	0.67	0.67
SSD	AraBERT	0.92	0.83	0.86	0.84
	DistilBERT	0.80	0.66	0.68	0.66
USD	AraBERT	0.92	0.84	0.84	0.84
	DistilBERT	0.81	0.67	0.69	0.67

4.4 Computational Efficiency

AraBERT consistently required longer training times than DistilBERT for both tasks. For topic classification, AraBERT averaged 18–19 min, while DistilBERT needed only 6–7 min. In sentiment classification, AraBERT's times ranged from 18 to 21 min, compared to DistilBERT's 6–8 min. The difference reflects AraBERT's higher model complexity.

5 Discussion

The experimental results reveal several key insights into the use of LLM-based pseudo-labelling and the broader landscape of Arabic ABSA in the healthcare domain. Notably, the findings affirm the efficacy of pseudo-labelling using ChatGPT, which proved to be a reliable annotator for Arabic texts. The minimal performance gap – often less than 0.02 in the F1 score – between models trained on machine-generated labels and those trained on fully human-labelled data underscores the potential of this method in low-resource environments. This is especially promising for languages or domains in which manually annotated corpora are scarce, suggesting that with well-designed prompting strategies and robust LLMs, it is possible to generate training data of sufficient quality to support high-performing classifiers.

Despite the strength of pseudo-labels, human reviews still add measurable value, particularly in refining fine-grained taxonomies. The FSD consistently yielded slightly better performance, as human annotators were able to catch subtle misclassifications – especially those requiring nuanced domain knowledge or deeper contextual understanding. This indicates that a hybrid approach, in which LLM-generated labels are supplemented by selective human validation, offers a cost-effective compromise. In fact, our semi-supervised experiments demonstrated that reviewing only 50% of the data can yield results comparable to full manual annotation, highlighting the diminishing returns of exhaustive human labelling when high-quality pseudo-labels are available.

Model selection also played a significant role, particularly in relation to Arabic-specific language modelling. AraBERT consistently outperformed DistilBERT across tasks, emphasising the advantages of using models tailored to the Arabic language.

Its deeper understanding of Arabic morphology, dialectal variations, and healthcare-specific terminology gave it a significant edge, especially in the more complex 17-class classification task. Nevertheless, the relatively strong performance of DistilBERT, despite being a general-purpose model, suggests its suitability for scenarios where slight trade-offs in accuracy are acceptable in exchange for faster training and inference times. This points to opportunities for future research to explore multilingual or domain-adapted models that strike a balance between performance and efficiency.

The impact of classification granularity was another important finding. Reducing the number of categories from 17 to 6 led to a notable improvement in classification performance. This supports the notion that the original taxonomy may have been too fine-grained or semantically overlapping, rendering consistent classification difficult. While detailed labels remain valuable for in-depth analysis, they also require more training data per class and are more prone to misclassification. A flexible taxonomy, as provided in the EHSAN dataset, allows researchers to select the level of detail appropriate for their specific objectives, and also encourages iterative taxonomy design – starting with detailed classes and clustering them based on confusion trends to achieve an optimal structure for machine learning applications.

Finally, a distinctive aspect of our dataset was the inclusion of model-generated rationales for each assigned label, aligning with the broader movement toward explainable AI in healthcare. Although we did not formally assess the impact of these explanations on user trust or debugging processes, qualitative feedback from the annotators suggested that they enhanced both confidence in the model's outputs and the interpretability of the annotation task. These rationales open up new possibilities, not only for enhancing annotation workflows but also for future model architectures that integrate explanation generation, such as multi-task learning setups or systems that warrant predictions alongside classification outputs.

Although the statistical analysis provided important insights into the reliability of the results, two considerations merit discussion. First, relying on only five runs may limit the precision of the confidence intervals, which could explain the noticeable variability observed in the confidence interval for the AraBERT model on the USD. In contrast, increasing the number of runs may lead to more stable estimates and narrower intervals. Second, the lack of statistical significance in some comparisons may be attributed to an underlying similarity in model behaviour across the different data settings, with only minor differences associated with the level of supervision. This is reflected in the comparison between the FSD and the USD in the AraBERT model, which did reach statistical significance.

In summary, the EHSAN framework proposed here offers a practical and scalable strategy for producing high-quality annotated datasets by effectively combining AI capabilities with human expertise. It enables fine-grained analysis of patient narratives in Arabic, demonstrating that, even in linguistically complex settings, modern NLP techniques can extract meaningful insights and support data-driven healthcare improvement. In terms of the study's limitations, considering that the present evaluation was limited to a specific healthcare domain and Saudi dialects, further work is needed to evaluate the framework's generalisability across other dialects and platforms. Another limitation is that our reliance on a single data source and language model may have introduced source- or model-specific biases.

6 Conclusion and Future Work

This study introduced EHSAN, a comprehensive Arabic dataset and framework for the ABSA of healthcare reviews. By leveraging ChatGPT for initial labelling and integrating human validation in a tiered manner, we addressed key limitations in prior Arabic healthcare NLP efforts: data scarcity, lack of aspect specificity, and opaque model decisions. Our results show that an Arabic-focused model (AraBERT) can achieve robust performance in classifying topics and sentiments in patient reviews, even with minimal human supervision, thereby affirming the viability of LLM-based pseudo-labelling for building reliable datasets. In practical terms, the EHSAN framework can be integrated into real-time digital feedback monitoring systems within hospitals, enabling the tracking of patient sentiment on specific aspects, such as waiting times or billing. These insights can support timely administrative actions and guide data-driven quality improvement initiatives.

Several avenues emerged for future work from this study. First, cross-hospital and cross-region generalisation should be examined. Our models could be tested on patient reviews from different countries or healthcare settings to evaluate how well the insights generalise and to possibly expand EHSAN with more diverse data. In particular, since this study focused on healthcare reviews from Saudi Arabia, it would be valuable to examine the applicability of the findings across other Arabic-speaking countries, especially those with distinct dialects and healthcare delivery models. Future research could evaluate model performance and explanation quality when applied to reviews written in different regional varieties of Arabic, thereby assessing the robustness and adaptability of the proposed framework across the wider Arab world. Second, enhancing the prompting strategy for ChatGPT could further improve annotation consistency and reduce the observed error rate. Third, the integration of explanations into model training is an exciting direction: a model could be trained to not only predict labels but also to generate justification, which would move us closer to truly interpretable AI in this context. Fourth, fairness evaluation represents another important direction. Language models may exhibit biases in their outputs. In this study, we incorporated human reviewers to validate the model's outputs and reduce the likelihood of unintended bias, particularly in the classification of aspects and sentiments. Therefore, future studies could include a more systematic assessment of model-generated rationales across different dialects and aspect categories, examining whether these rationales contain implicit or unwarranted assumptions. In addition, analysing classification errors could help uncover potential bias patterns and address them through more precise and targeted prompting strategies to improve model fairness across varying contexts. Finally, enhancing dataset quality, such as increasing the number of annotated samples and ensuring balanced class distributions, may help reduce performance variability and lead to more statistically significant results across training configurations.

From a data-management standpoint, EHSAN already delivers a solid, production-ready pipeline; the next step is to amplify its reach. Three opportunities stand out for the

IDEAS community to extend the platform's impact even further: 1) Streaming schema evolution: The aspect taxonomy must grow gracefully as hospitals introduce new services (e.g., a tele-ICU programme / remote intensive-care monitoring over video links). Supporting such additions on the fly, without interrupting dashboards or historical queries, will keep insights continuously up to date. 2) Explanation-provenance graphs: For every sentiment label the system assigns, it should be possible, instantly, to trace a path back to the exact ChatGPT prompt, the model's raw reply and any human corrections. Storing and querying this chain of evidence is non-trivial at scale. 3) Privacy-aware indexing: When we let users retrieve "reviews most similar to this one," the search index must guarantee k-anonymity so that rare disease mentions or unique patient details cannot be used to re-identify individuals.

In conclusion, the EHSAN dataset and approach fill a critical gap in Arabic healthcare text analysis by providing a high-resolution, explainable view of the patient experience. We hope that this work lays the groundwork for more patient-centric analytics in Arabic and other low-resource languages, and that the methods presented will inspire further innovations at the intersection of human expertise and AI-powered language understanding.

An anonymised version of the EHSAN dataset and the experimental code has been archived on Zenodo for perpetual access (https://doi.org/10.5281/zenodo.15418860).

References

1. Greaves, F., Ramirez-Cano, D., Millett, C., Darzi, A., Donaldson, L.: Use of sentiment analysis for capturing patient experience from free-text comments posted
2. Zunic, A., Corcoran, P., Spasic, I.: Sentiment analysis in health and well-being: Systematic review. JMIR Med. Inform. **8**(1), e16023 (2020)
3. Ferreira, D.C., Vieira, I., Pedro, M.I., Caldas, P., Varela, M.: Patient satisfaction with healthcare services and the techniques used for its assessment: a systematic literature review and a bibliometric analysis. Healthcare **11**, 639 (2023)
4. AlNasser, S., AlMuhaideb, S.: Listening to patients: advanced arabic aspect-based sentiment analysis using transformer models towards better healthcare. Big Data Cogn. Comput. **8**(156), 1–27 (2024)
5. Lumeon, 90% of patient access leaders rank patient experience as the key differentiator for hospitals. https://info.lumeon.com/patient-access-leadership-research-report. Accessed 10 April 2025
6. CMS: HCAHPS Fact Sheet (CAHPS ® Hospital Survey) (2022)
7. Ministry of Health, Saudi Arabia. https://www.moh.gov.sa/en/Pages/Default.aspx. Accessed 10 April 2025
8. Beattie, M., Murphy, D.J., Atherton, I., Lauder, W.: Instruments to measure patient experience of healthcare quality in hospitals: a systematic review. Syst. Rev. **4**(97), 1–21 (2015)
9. Osváth, M., Yang, Z.G., Kósa, K.: Analyzing narratives of patient experiences: a BERT topic modeling approach. Acta Polytechnica Hungarica **20**(7), 153–171 (2023)
10. Bansal, A., Kumar, N.: Aspect-based sentiment analysis using attribute extraction of hospital reviews. New Gener Comput. **40**, 941–960 (2022)
11. Alhazzani, N.Z., Al-Turaiki, I.M., Alkhodair, S.A.: Text classification of patient experience comments in Saudi dialect using deep learning techniques. Appl. Sci. (Switzerland) **13**(10305), 1–27 (2023)

12. Malik, U., Bernard, S., Pauchet, A., Chatelain, C., Picot-Clemente, R., Cortinovis, J.: Pseudo-labeling with large language models for multi-label emotion classification of French tweets. IEEE Access **12**, 15902–15916 (2024)
13. Li, J., et al.: Identifying healthcare needs with patient experience reviews using ChatGPT. PLoS ONE **20**(3), e0313442 (2025)
14. Al-Thubaity, A., et al.: Evaluating ChatGPT and Bard AI on Arabic sentiment analysis. In: Proceedings of ArabicNLP, pp. 335–349. Association for Computational Linguistics, Singapore (2023)
15. Saudi Central Board for Accreditation of Healthcare Institutions. Hospitals accreditation status till 1 May 2025 (2025). https://portal.cbahi.gov.sa/Library/Assets/hos-ar-035644.pdf. Accessed 10 April 2025
16. Qi, P., Zhang, Y., Zhang, Y., Bolton, J., Manning, C.D.: Sta n z a : a Python natural language processing toolkit for many human languages. In: Proceedings of the 58th Annual Meeting of the Association for Computational Linguistics: System Demonstrations, pp. 101–108, Association for Computational Linguistics (2020)
17. Cohen, J.: A coefficient of agreement for nominal scales. Educ. Psychol. Measur. **20**(1), 37–46 (1960)
18. Antoun, W., Baly, F., Hajj, H.: AraBERT: transformer-based Model for Arabic language understanding. In: Proceedings of the 4th Workshop on Open-Source Arabic Corpora and Processing Tools, with a Shared Task on Offensive Language Detection, pp. 9–15. European Language Resource Association, Marseille (2020)
19. Sanh, V., Debut, L., Chaumond, J., Wolf, T.: DistilBERT, a distilled version of BERT: smaller, faster, cheaper and lighter (2019)

Automated Glyph Feature Detection Using Convolutional Neural Networks

Michael Mason(✉)[iD], Sam Kirchner[iD], and Carter Powell[iD]

University of Nebraska – Lincoln, Lincoln, NE 68588, USA
mmason12@unl.edu, {skirchner3,cpowell7}@huskers.unl.edu

Abstract. A glyph is a specific design for a character in a writing system. Analyzing a glyph's anatomical features can offer insight into its applications, ancestry, and historical context. However, manually identifying features is a subjective, time-consuming task. In this paper we present the Automated Letter Feature Analyzer (ALFA) system for computationally identifying a glyph's anatomical features. ALFA uses convolutional neural networks (CNNs) along with other image analysis techniques to evaluate glyphs using both learned patterns and explicit shape metrics. A modular web-based framework was created to efficiently render, capture, and label large glyph image datasets for machine learning tasks. CNNs were trained to detect three anatomical features with overall accuracies between 97% and 99%. The system also achieved an accuracy of 98.45% when counting enclosed spaces and objects in positive space, while glyph weight by quadrant was tuned to concur with visual labeling at 97% accuracy. Results show ALFA is not only useful for collecting glyph images and labeling large image datasets but also can facilitate new research in computational linguistics by offering a way to computationally detect a glyph's anatomical elements.

Keywords: Optical character recognition · Typography · Convolutional neural networks · Feature analysis

1 Introduction

Wherever the written word exists, it is accompanied by intentional and unintentional choices about how the signs should be represented. These choices culminate in a glyph; a specific design for some character in a writing system. Glyphs are aggregated into typefaces, also called font families, based on their anatomical features and overall design [5]. Analyzing a glyph's anatomical features can offer insight into its ancestry, applications, and historical context [3]. However, manually labeling features is a subjective, labor-intensive process often requiring expert knowledge.

To simplify the feature labeling process, we introduce the Automated Letter Feature Analyzer (ALFA), a web-based suite of tools for amassing glyph images, annotating image datasets, and computationally generating glyph feature vectors. ALFA employs convolutional neural networks (CNNs) to detect abstract

© The Author(s), under exclusive license to Springer Nature Switzerland AG 2026
https://doi.org/10.1007/978-3-032-06744-9_3

features based on learned patterns. In addition, ALFA uses more predictable methods to extract structural information like how the glyph's pixels are distributed geographically. The system's feature detection capabilities were tested using a large synthetic dataset composed of Unicode characters rendered in multiple typefaces. The image dataset was annotated using a custom-built labeling interface included on the website. A public version of the ALFA site is under active development, and can be found at https://alfa.unl.edu.

The rest of the paper is divided as follows. Section 2 reviews previous work on computationally analyzing glyph features. Section 3 outlines our process for collecting and labeling a large number of glyph images. Section 4 describes our CNN architectures and other analysis methods. Experimental results are presented in Sect. 5, then discussed in Sect. 6. Finally, Sect. 7 summarizes our work, outlines planned improvements, and offers ideas for future research.

2 Related Work

As part of showing the relationship of scripts within the Cretan script family, Revesz [13] introduced a method for comparing glyphs based on vectors of 13 Boolean values where each value represents the presence or absence of a specific anatomical feature. Checking the number of shared values between vectors provides a numerical measure of similarity between glyphs. This method can be extended to create feature matrices for measuring the similarity between writing systems. Using this approach, Revesz was able to reconstruct a phonetic grid for the undeciphered Linear A script based on sounds from known scripts closely related to Linear A. The phonetic grid was used to provide convincing translations of twenty-eight Linear A inscriptions and one Etocretan inscription, demonstrating script comparison via feature vectors holds merit as a general-purpose tool for deciphering unknown scripts. This work was expanded in Alamuru et al. [2] by adding six additional features to measure the similarity between the Indus Valley script and modern Dravidian scripts like Tamil and Telugu. Beyond the pursuit of deciphering ancient scripts, Chu et al. [7] present a method for decomposing Chinese glyphs into meaningless base components for the purpose of secret sharing via visual cryptography.

New advancements in glyph analysis are supported by contemporary computer vision techniques and the rapid rise of convolutional neural networks. CNNs facilitate glyph research which was not previously possible because they can be trained to inspect images for subtle patterns imperceptible to humans, and can quickly make an enormous number of comparisons between glyphs that would take a human decades to do by hand. This was shown in work by Daggumati and Revesz, who applied CNN-based models to investigate the historical transmission of Bronze Age writing systems [10] and mined structural relationships between scripts using similarity-driven algorithms [9].

3 Dataset Engineering Process

3.1 Glyph Curation

CNNs typically require thousands of labeled images to train. Rather than draw thousands of glyphs by hand, we can construct datasets from the wide variety of symbols offered by Unicode. The modern Unicode Standard 16.0 supports 154,998 characters across 168 scripts both ancient and modern [8]. For our purposes, we assume each Unicode character is unique.

A webpage was created to render select Unicode blocks (shown in Fig. 1). This script display page was later merged with other tools to form the ALFA website. Each script is supported by at least one typeface, but many scripts are displayed using multiple typefaces to increase the number of visually distinct glyphs in our dataset. Table 1 lists the 48 scripts used to populate the display page and their corresponding typefaces. Georgian is a special case; its Unicode block contains multiple typefaces by default. A minimum font size of 50 pixels was used to maintain an image fidelity comparable to previous publications using CNNs for script analysis [9,10]. Most diacritics unaccompanied by a base glyph were removed from the dataset for redundancy and perceived lack of complexity.

After populating the display page, the Selenium WebDriver browser automation library [14] was used to capture and save each glyph on the page as a .png file. Each typeface was assigned a square bounding box of 65, 75, or 90 pixels wide to accommodate its largest glyph. Using this process on the listed scripts and typefaces, we obtained 8,497 images.

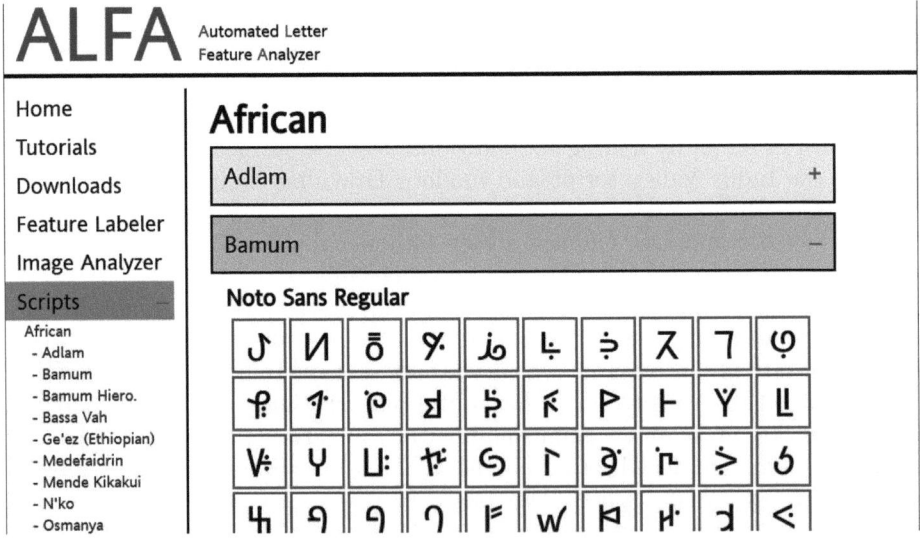

Fig. 1. A screenshot of the script display page. Scripts are separated into families and may include multiple typefaces.

Table 1. Scripts used and their corresponding typefaces.

Typeface	Script
Aegean	Carian, Cretan Hieroglyphs, Cypro-Minoan, Cypriot, Linear A, Linear B Ideograms, Linear B Syllabary, Lycian, Lydian, Phaistos Disk.
BabelStone Special	Ogham.
GNU FreeMono	Anglo-Saxon Runes, Dalecarlian Runes, Elder Futhark, Medieval Runes, Younger Futhark (long-branch), Younger Futhark (short-twig).
GNU FreeSans	Vai.
Junicode-Ansund	Anglo-Saxon Runes, Dalecarlian Runes, Elder Futhark, Medieval Runes, Younger Futhark (long-branch), Younger Futhark (short-twig).
Noto Sans	Adlam, Anatolian Hieroglyphs, Anglo-Saxon Runes, Armenian, Bamum, Bamum Hieroglyphs, Bassa Vah, Carian, Caucasian Albanian, Cherokee, Cypro-Minoan, Cypriot, Dalecarlian Runes, Elbasan, Elder Futhark, Ge'ez (Ethiopian), Georgian, Kaktovik Numerals, Linear A, Linear B Ideograms, Linear B Syllabary, Lycian, Lydian, Medefaidrin, Medieval Runes, Mende Kikakui, Nag Mundari, N'ko, Ogham, Ol Chiki, Old Hungarian, Old Turkic, Osage, Osmanya, Pahawh, Sora Sompeng, Tifinagh, Unified Canadian Aboriginal Syllabics, Vai, Vithkuqi, Warang Citi, Younger Futhark (long-branch), Younger Futhark (short-twig).
Noto Sans Symbols 2	Alchemical Symbols.
Noto Serif	Ge'ez (Ethiopian), Nyiakeng Puachue, Todhri, Vithkuqi.
Symbola	Alchemical Symbols.
Times New Roman	Cypriot.

3.2 Feature Classification

Each feature requires a separate CNN for detection, and each CNN requires a dataset of labeled images for training. As a proof-of-concept, we implemented models to detect three glyph features used in previous work [2,12]: a "wedge" on top, curved lines, and a vertical midline. Definitions for these features are now described in detail.

We borrow a partial definition from the field of typography to redefine an upper wedge as an apex; a place where the glyph's highest point is also the meeting place of two strokes [5]. Hence, a glyph has an apex if the vertex of an angle has the same y-coordinate as the topmost pixel in the glyph. If a ray of the apex angle is a flat horizontal line as in the capital English letter E, it is not considered a wedge. Defining a curved line became an unexpected challenge. Some writing implements can produce curved edges despite only being used to draw straight lines because of their tip's geometry or the material properties of the writing medium and implement. Thus, we define a line as curved if it cannot be drawn using only straight strokes. Dots of any size, despite being circular, are excluded from the definition of a curved line because they do not have two points. A "vertical midline" is defined as a vertical line running directly through the center of the glyph whose highest and lowest points are equal to the highest and lowest parts of the glyph.

There were many examples where symbols did not fit cleanly into a feature's definition but were close enough multiple researchers considered the glyph to be an acceptable exception to the definition. These issues were address on a case-by-case basis, and are a primary motivator for attaching percentile confidence values to each true/false label the CNNs assign.

Manually labeling thousands of images for machine learning tasks is a tedious, time consuming process. To expedite and simplify the process, an image labeling tool was created and integrated into the ALFA website (Fig. 2). The feature labeler supports both true/false and integer labeling modes. A second round of labeling was conducted after preliminary system testing found classification errors made by the human labelers.

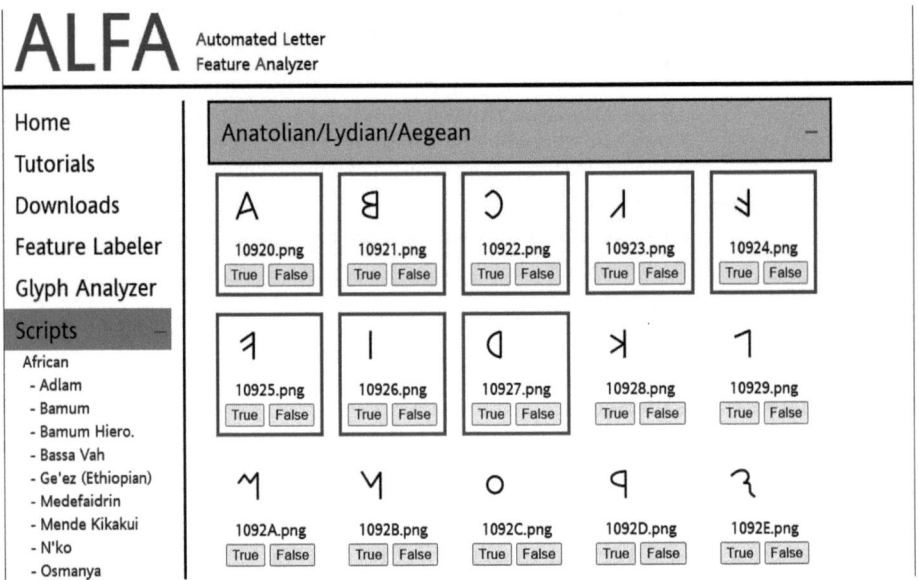

Fig. 2. A screenshot of the feature labeler page in true/false mode. Users are able to upload images then use the graphical interface to efficiently annotate machine learning datasets.

3.3 Dataset Analysis

After initial labeling, each feature-specific dataset underwent preprocessing and augmentation to improve class balance and training coverage. All images were cropped to remove excess whitespace, converted to grayscale, and resized to

64 × 64 pixels. Each dataset was augmented to increase its size, then randomly partitioned using an 80:20 train-test split after augmentation.

Augmentation strategies were tailored for each feature to preserve label integrity. The apex dataset was augmented using Gaussian noise. No rotations were applied because the feature is sensitive to changes in glyph orientation. The apex dataset's original size of 8,080 images was increased to a final size of 25,856. The curved line dataset was augmented by rotating each image by 90°, 180°, and 270° to exploit orientation insensitivity, quadrupling the size of the original dataset. The vertical midline dataset was augmented by applying a 180° rotation to exploit the feature's rotational symmetry, doubling the size of the original dataset. Additional details can be seen in Table 2.

Table 2. Final Dataset Information

Model	Class	Final Count	Total	Class Imbalance (T:F)
Apex	True	2,396	25,856	1 : 9.791
	False	23,460		
Curved Line	True	18,060	25,700	2.364 : 1
	False	7,640		
Vertical Midline	True	1,522	16,992	1 : 10.164
	False	15,470		

4 System Architecture

ALFA runs convolutional neural networks in parallel with a secondary image analysis pipeline to evaluate glyphs using both learned patterns and explicit shape metrics. This hybrid design merges the flexibility of deep learning with the stability of more straightforward, predictable methods.

4.1 Convolutional Neural Networks

Primary feature detection is done using convolutional neural networks to search for abstract patterns in glyphs. In addition to a true/false classification indicating the feature's presence, ALFA returns a percentile confidence value indicating how strongly the system believes the classification is correct. Augmenting binary feature vectors with confidence values enables more nuanced vector comparisons than previously shown, and helps compensate for errors caused by subjectivity during the labeling process.

All components of the CNNs are implemented in Python using the TensorFlow [1] and OpenCV [4] libraries. Each greyscale image is normalized to the $[0, 1]$ range. The models are defined using the Keras Sequential API [6], and consist of stacked Conv2D layers with ReLU activation, followed by BatchNormalization, MaxPooling2D, and optional dropout for regularization. Each network ends with a sigmoid neuron to return a binary classification with a percentile confidence value. This standard architecture is based on a publicly available CNN implementation for the Fashion MNIST dataset [11]. LeakyReLU and PReLU were both tested as activation functions, but achieved poorer results. L2 weight regularization is applied to selected convolutional and dense layers to mitigate overfitting. The networks are compiled with the Adam optimizer and trained using binary cross-entropy loss. Early stopping and model checkpointing are used to monitor validation performance and preserve the best models. Class weights are computed dynamically to address class imbalances. Evaluation includes accuracy, precision, recall, and F1-score using scikit-learn's confusion matrix and classification report.

We implemented three distinct CNN architectures to detect different glyph features: an apex, curved line, and vertical midline. The apex model does not use L2 regularization, but applies stronger dropout to reduce overfitting. It begins with convolutional blocks using 128, 256, and 512 filters, each followed by BatchNormalization and MaxPooling. After flattening, a dense layer with 128 units and ReLU activation is used, followed by a dropout layer with rate 0.6. The curved line model uses L2 regularization with $\lambda = 0.05$ to prevent overfitting. The network consists of three convolutional blocks with 64, 128, and 256 filters, respectively. Each block includes a Conv2D layer with ReLU activation, followed by BatchNormalization and a 2×2 MaxPooling layer. After flattening, a dense layer with 64 units and ReLU activation is applied, followed by a dropout layer with rate 0.4. The vertical midline model uses the same architecture as the curved line model, but with fewer filters and a smaller L2 regularization factor of $\lambda = 0.005$. The convolutional blocks use 32, 64, and 128 filters, followed by BatchNormalization and MaxPooling. The fully connected portion includes a dense layer with 64 units and dropout with rate 0.4.

4.2 Glyph Decomposition

ALFA expands upon previous publications [2,12] using a Boolean "enclosed space" feature by replacing a binary data type with a numerical value. To achieve this, each glyph is decomposed into a set of disconnected objects and enclosed counters by comparing where negative and positive space meet based on the method presented in Chu et al. [7]. An enclosed counter is the phrase used in typography to describe an area of negative space completely enclosed by an object in positive space [5]. In this context, positive space refers to all elements used to draw the glyph, while negative space refers to the background a glyph is rendered onto.

Glyph decomposition begins with preprocessing. Images are upscaled from their original size to an 800×800 pixel square using Lanczos interpolation. Lanczos interpolation is more computationally intensive than simple linear interpolation, but produces smoother results because it calculates the interpolated pixel's weight from all eight adjacent pixels instead of just four from the cardinal directions. A small amount of Gaussian blur is applied within a five pixel range of every interpolated pixel to smooth any rough edges created during the interpolation process.

The greyscale image is converted to a black-and-white matrix, and a set of contour lines are created by tracking where black and white pixels meet. These contour lines are added to a tree-based structure to count how many transitions between clusters of black and white pixels have occurred. A contour's level in the tree is used to determine if it represents a object or enclosed counter.

4.3 Glyph Balance

We calculate three additional non-Boolean metrics based on the number of pixels in each of quadrant of the glyph. These metrics are top/bottom weight, left/right weight, and dominant quadrant. A glyph is either top/bottom (left/right) heavy or balanced. Similarly, a glyph is either balanced or has a dominant quadrant. These weight metrics offer an elementary measure of symmetry, and provide useful information when identifying glyphs. For example, the English letters p and d are isomorphic if compared using the three true/false features and glyph decomposition metrics presented thus far, but can still be differentiated by their top/bottom weights.

Balance calculations begin by cropping the image to exclude excess whitespace surrounding the glyph. The image is again converted to black and white, but is not upscaled or downscaled to avoid altering the glyph's weights. The glyph is divided along its vertical and horizontal midlines into four equal quadrants. Each quadrant's pixels are counted, then the tallies are compared to calculate the three balance metrics. A balance threshold is introduced to account for small, insignificant differences not visible to the human eye. Too large of a threshold risks marking unbalanced symbols as balanced, and too small of a threshold can lead to unexpected results where clearly balanced symbols are marked as biased towards one side due to artifacts introduced by things like storing the image using lossy compression or quantization decisions made while rastering a Bezier curve to a pixel map. Quadrants are considered balanced if they fall within 5% of each other, and halves are considered balanced if they fall within 12.5%.

5 Results

5.1 Convolutional Neural Networks

All models were trained to analyze a 64×64 grayscale image. Evaluation was conducted on a held-out test set. All work with CNNs was done on a consumer-grade Windows 10 desktop computer with an RTX 4070 TI graphics card, AMD Ryzen 7900X3D processor, and 32 GB of DDR4 RAM. Confusion matrices and graphs depicting model accuracy plotted against training epoch for the apex, curved line, and vertical midline models are shown in Figs. 3, 4, and 5, respectively. CNN performance metrics are summarized in Table 3.

The apex model achieved an overall accuracy of 98%. The "False" class was assigned with a precision of 0.99 and recall of 0.99, while the "True" class achieved a precision of 0.90 and recall of 0.87. The training and validation accuracy curves indicate fast convergence within the first 10 epochs, with both stabilizing above 98%. The relatively small distance between training and validation curves indicates there is little overfitting. The confusion matrix shows 1433 true negatives, 145 true positives, 17 false positives, and 21 false negatives. Mean and median training times across five runs are 378.3 and 446 s, respectively.

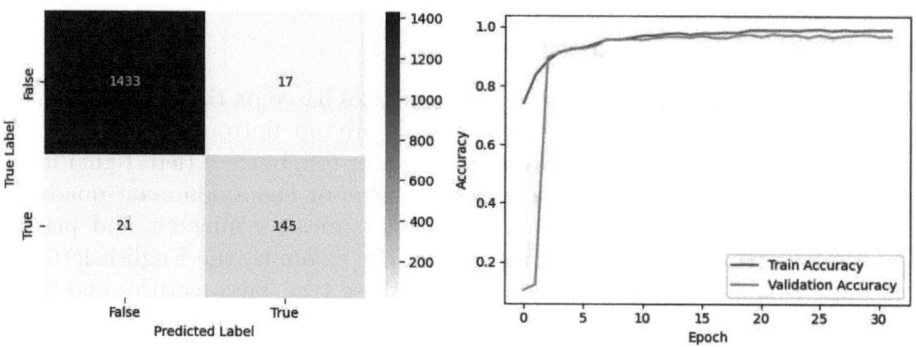

Fig. 3. Confusion matrix (left) and model accuracy (right) for the apex model.

The curved line model achieved an overall accuracy of 97%. The "False" class was assigned with a precision of 0.95 and recall of 0.97, while the "True" class achieved a precision of 0.99 and recall of 0.98. Although validation accuracy is more volatile, both curves show rapid convergence within the first 20 epochs, after which performance stabilized above 97%. These curves are relatively close, which indicates there is little overfitting. The confusion matrix shows 1505 true negatives, 3505 true positives, 48 false positives, and 82 false negatives. Mean and median training times across five runs are 512.6 and 412.7 s, respectively.

The vertical midline model achieved an overall accuracy of 99%. The "False" class was assigned with a precision of 0.99 and recall of 0.99, while the "True" class achieved a precision of 0.93 and recall of 0.94. The training and validation accuracy curves show rapid convergence within the first 10 epochs, with performance stabilizing near 99%. The training and validation curves are well-aligned, indicating there is little overfitting. The confusion matrix shows 2319 true negatives, 219 true positives, 17 false positives, and 15 false negatives. Mean and median training times across five runs are 92.98 and 77.3 s, respectively.

5.2 Glyph Decomposition

Decomposition was tested on 773 glyphs annotated using ALFA's feature labeling page. Glyphs were selected to include both simple and complex examples from

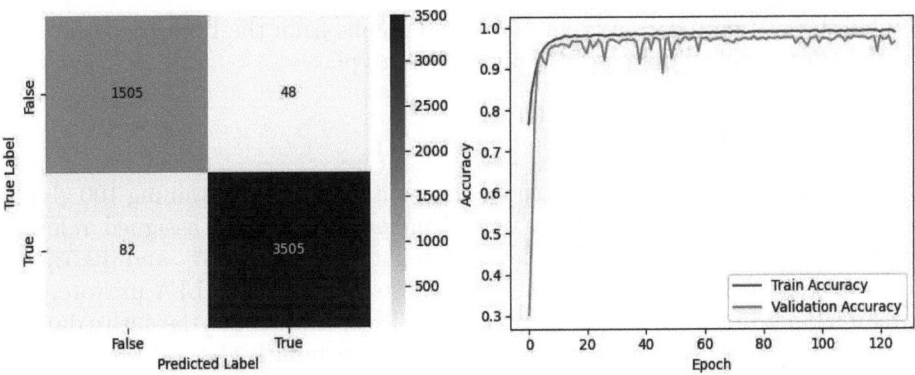

Fig. 4. Confusion matrix (left) and model accuracy (right) for the curved line model.

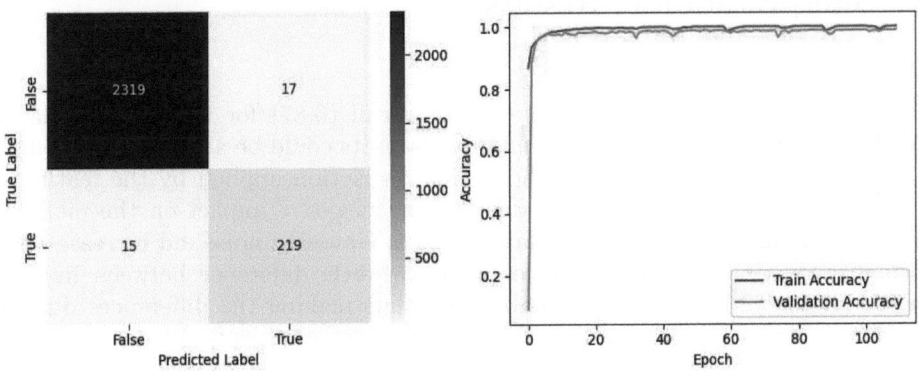

Fig. 5. Confusion matrix (left) and model accuracy (right) for the vertical midline model.

Table 3. A summary of performance metrics for each CNN. Mean and median training times across five runs are in seconds.

Model	Class	Precision	Recall	F1-score	Accuracy	Mean Time	Median Time
Apex	True	0.90	0.87	0.88	0.98	378.3	446
	False	0.99	0.99	0.99			
Curved Line	True	0.99	0.98	0.98	0.97	512.6	421.7
	False	0.95	0.97	0.96			
Vertical Midline	True	0.93	0.94	0.93	0.99	92.98	77.3
	False	0.99	0.99	0.99			

every script. A consumer-grade desktop computer with 16 GB of DDR4 RAM and a Ryzen 7 1700 processor analyzed the set of 773 images in approximately 105 s, scoring an accuracy of 98.45% for both decomposition metrics. Of the 773 glyphs, the system incorrectly analyzed two glyphs from the Phaistos Disk, nine Cretan Hieroglyphs, and one Anatolian Hieroglyph.

5.3 Glyph Balance

Glyph balance thresholds were tuned using a small dataset containing 100 glyphs of mixed complexity. Each glyph was visually assessed and assigned relevant labels. Best test results were achieved using thresholds of 5% and 12.5% for quadrants and halves, respectively. Using these parameters, ALFA mirrored the human labeler in 97% of weight comparisons. Total runtime for the entire dataset of 8,497 glyphs on the same machine used to benchmark glyph decomposition was approximately five seconds.

6 Interpretation of Results

6.1 Convolutional Neural Networks

Our apex detection model had the lowest recall (0.87) for the positive class. While the curved line and vertical midline datasets could be augmented through rotation, the apex dataset could not. This restriction applied by the feature's orientation sensitivity may have had a strong negative impact on the model's recall. Although augmenting the dataset using Gaussian noise did increase performance, applying noise introduces a more subtle difference between images in the augmented dataset than a rotation, thus making the differences during training less useful for the system. Representative examples of false positives and negatives can be seen in Fig. 6.

Fig. 6. From left to right, representative examples of three false positives and two false negatives for an apex. False positives were caused by wedges being not quite on top, disconnected strokes, or curves misinterpreted as wedges. False negatives were caused by thick lines and sharp, pointed curves being misinterpreted as flat or strongly curved.

Our model for detecting curved lines performed the strongest with a recall and F1-score of 0.98 for the positive class. Its performance over the other models may be explained by the dataset's more symmetric class imbalance of 2.364:1 contrasted with the ratios 1:9.791 and 1:10.164 found elsewhere. Furthermore, extensive use of rotation-based augmentation to exploit this feature's indifference to orientation may have created a more robust dataset by introducing images with greater meaningful differences than the augmentation techniques used for the other features. Dots were excluded from the definition of a curved line, but sufficiently large dots resulted in multiple false positives by emulating curved lines. Similarly, straight lines with blurry edges were likely interpreted as curves because of artifacts introduced through scaling and anti-aliasing. Many false negatives display bold, clearly straight lines that overshadow a much smaller curved segment. This indicates the system may have developed heuristics like assuming there are no curved lines in the glyph if it first examines an area with strong, straight lines. Examples of these errors can be seen in Fig. 7.

Fig. 7. Representative examples of false positives and negatives for curved lines. The glyph on the right is likely classified as curved despite not meeting our technical definition because it is composed of three dots. Artifacts created by anti-aliasing and upscaling have given the partial glyph on the left a wavy edge, throwing another false positive. The middle glyph is classified as not curved despite having a circle, perhaps because the circle is dominated by the more prominent straight lines.

Our model for detecting a vertical midline also performed well with a recall of 0.94 for the positive class. The dataset for this task was significantly imbalanced, with a roughly 10:1 ratio between the negative and positive classes. Despite this limitations, the model maintained low false positive and false negative rates, indicating effective learning under both augmentation and imbalance constraints. Examples of false positives and negatives can be seen in Fig. 8.

Fig. 8. From left to right, representative examples of three false positives and two false negatives for a vertical midline. False positives were caused by midlines which curve at the end, vertical lines not actually in the center, and midlines which do not go all the way from top to bottom.

6.2 Glyph Decomposition

The close relationship between objects and enclosed counters almost guarantees the two tallies are either both right or both wrong. Decomposition performed well overall with an accuracy above 98%, but the system had difficulty processing glyphs with small, complex features or a low-quality base image. Artifacts introduced during the upscaling process occasionally broke existing lines or merged objects. Some of these issues can be addressed by tuning what greyscale values are mapped to black or white during the conversion process; previous accuracy results were improved from 95% to their current level by expanding the range of what greyscale values are mapped to black by 2.73%. Implementing more advanced interpolation techniques my also help address line continuity issues.

6.3 Glyph Balance

Our method for computing balance metrics is assumed to always be computationally correct even if it disagrees with testing data labeled by humans. Glyphs intended to appear symmetrical are sometimes designed with anatomically asymmetrical components to accommodate for visual heuristics and optical illusions [5]. Thus, a human's labels should not be expected to perfectly overlap with those given by the system. It could be argued balance thresholds are not appropriate because the system is designed to process images in an objective way untainted by human error. However, balance with and without a threshold provide different information. Tuning balance thresholds to closely match a human labeler allows the system to simulate how a human might interpret a glyph's balance. A comparison between simulated human balance metrics and purely objective ones could provide type designers with a fast, evidence-based way to evaluate the perceptual impact of certain design changes.

7 Conclusions and Future Work

Preliminary results show ALFA is not only useful for collecting glyph images and labeling large image datasets, but is capable of being trained to detect glyph features with a high overall accuracy. Incorporating confidence-weighted true/false and numerical values to create more wholistic feature vectors enables more nuanced approaches to similarity measures than previously available. In addition to enabling more advanced comparison methods between glyphs, the easy computation of feature vectors facilitates more accurate and expansive script comparisons not previously possible due to scale and requisite knowledge.

Although only three CNNs were included in this version of ALFA, the system can be reasonably trained to detect almost any feature if given a sufficiently large and accurately labeled dataset. The system could be improved by writing code to crop, scale, and compare symbols to filter out duplicate glyphs. Adding more scripts and typefaces to expand the dataset and implementing a consensus protocol to have multiple architectures vote on the existence of a feature should improve accuracy. The ALFA website is under active development, and we intend to soon finalize pages for users to download assets and process images.

Disclosure of Interests. The authors have no competing interests to declare that are relevant to the content of this article.

References

1. Abadi, M., et al.: Tensorflow: a system for large-scale machine learning. In: Proceedings of the 12th USENIX Symposium on Operating Systems Design and Implementation (OSDI 2016), pp. 265–283 (2016). https://www.tensorflow.org/
2. Alamuru, S.S.S., Barla, S.S., Revesz, P.Z.: Feature analysis of indus valley and dravidian language scripts with similarity matrices. In: Proceedings of the 26th International Database Engineered Applications Symposium, New York, NY, USA, pp. 63–69. IDEAS '22, Association for Computing Machinery (2022)
3. Berry, J.D.: Language Culture Type: International Type Design in the Age of Unicode. ATypI, New York and N.Y. (2002)
4. Bradski, G.: The OpenCV Library. https://opencv.org/
5. Carter, R., Day, B., Maxa, S., Meggs, P.B., Sanders, M.: Typographic Design: Form and Communication. Wiley-Blackwell, New York (2014)
6. Chollet, F.: Keras: the Python deep learning library. https://keras.io/
7. Chu, H., Huang, T., Kang, L., Lin, C., Wang, Y.: A novel non-negative matrix factorization technique for decomposition of Chinese characters with application to secret sharing. In: EURASIP Journal on Advances in Signal Processing (2019)
8. Consortium, T.U.: The Unicode Standard. https://www.unicode.org/versions/Unicode15.1.0/
9. Daggumati, S., Revesz, P.: Data mining ancient scripts to investigate their relationships and origins. In: Proceedings of the 23rd International Database Engineering & Applications Symposium (IDEAS 2019), pp. 1–10 (2019). https://doi.org/10.1145/3331076.3331116
10. Daggumati, S., Revesz, P.Z.: Convolutional neural networks analysis reveals three possible sources of bronze age writings between Greece and India. In: Information (2023). https://www.mdpi.com/2078-2489/14/4/227
11. Fleming, E.: Fashion MNIST Analysis and Visualization. http://www.eamonfleming.com/projects/fashion-mnist.html
12. Revesz, P.: Establishing the west-ugric language family with Minoan, Hattic and Hungarian by a decipherment of linear a. In: WSEAS Transactions on Information Science and Applications, pp. 306–335 (2020). https://wseas.com/journals/isa/2017/a605909-068.pdf
13. Revesz, P.Z.: Bioinformatics evolutionary tree algorithms reveal the history of the Cretan Script Family. International Journal of Applied Mathematics and Informatics **10**(1), 67–76 (2016)
14. Selenium: Selenium WebDriver (2024). https://www.selenium.dev/documentation/webdriver/

A Vision for Robust and Human-Centric LLM-Based QR Code Security

Hissah Almousa(✉) and Ellis Solaiman

School of Computing, Newcastle University, Newcastle upon Tyne NE1 7RU, UK
{h.almousa2,ellis.solaiman}@newcastle.ac.uk

Abstract. Quick Response (QR) codes are now widely used as a digital communication tool. However, their extensive adoption has made them an attractive target for cyberattacks, particularly through the injection of malicious URLs that redirect users to phishing sites or initiate malware installations. Conventional security approaches such as blacklists and antivirus software are no longer efficient against such evolving threats. This vision paper proposes an AI-based framework using fine-tuned Large Language Models (LLMs) to identify malicious URLs embedded within QR codes. To ensure transparency, a novel ensemble Explainable AI (XAI) is applied to aggregate insights from various XAI methods to explain the features influencing model predictions, facilitating more robust interpretations. To enhance clarity and usability, the proposed framework incorporates personalized explanations tailored to cybersecurity analysts, system developers, and non-expert end users, informed by a role-specific user study. Furthermore, as XAI methods may expose sensitive model behavior, cyberattackers craft adversarial inputs to mislead the model or manipulate explanations. This necessitates the integration of adversarial training to ensure model robustness and explanation integrity evaluated through perturbation consistency checks. The paper outlines key challenges in explanation fidelity and personalization and presents a development roadmap to advance secure, transparent, and human-centric explainable QR code analysis.

Keywords: QR Code Security · Large Language Models (LLMs) · Malicious URL Detection · Explainable AI (XAI) · Personalized Explanations · Adversarial Robustness

1 Introduction

According to QR Code Statistics, in 2025, global QR code scans are estimated to exceed 41 million, representing a 433% increase in just four years. QR codes have become widely adopted for mobile payments, marketing, and information

---

This is submitted as a vision paper to IDEAS 2025, outlining a research agenda and proposed framework to be developed and validated in ongoing work.

access, particularly with 54.33% linked to URL-based services. However, this widespread use has made them vulnerable to cyberattacks, where attackers exploit QR codes to inject harmful URLs for malicious purposes. Traditional defenses to detecting malicious URLs, including antivirus software, URL blacklist, and classical Machine Learning (ML) models, are increasingly falling short against today's advanced threats [17]. While these solutions provide some level of protection, the expanding nature of adversarial tactics presents a growing challenge. Additionally, these models are often treated as black boxes, offering little to no transparency about their decision-making processes. In high-stakes domains such as cybersecurity, this lack of interpretability undermines user trust, hinders accountability, and limits practical adoption. Moreover, adversaries increasingly exploit this ambiguity to craft inputs or manipulate model behavior in ways that bypass detection or compromise explanation systems [15].

Explainable Artificial Intelligence (XAI) has emerged as a promising solution to bridge the gap between performance and trust. The General Data Protection Regulation (GDPR) law in Europe now requires that meaningful explanations must be provided to people who are affected by automated decision-making systems. By providing human-understandable justifications for AI predictions, XAI offers a pathway toward more transparent and trustworthy decision-making. Meanwhile, recent advances in Large Language Models (LLMs), such as Transformers, have shown remarkable success in learning the deep semantic patterns of textual data [14], including URLs. However, these advancements have yet to be fully leveraged for QR code security in a way that integrates explainability, robustness, and usability.

Existing systems either lack explainability, are not adversarially robust, or fail to adapt explanations to diverse users. Consequently, this vision paper proposes a novel and forward-looking direction: the development of a secure, explainable, and adversarially robust AI-based defense system for detecting malicious URLs embedded within QR codes. The proposed approach combines fine-tuned LLMs for accurate classification with adaptive XAI methods, such as SHapley Additive exPlanations (SHAP) [9] and Local Interpretable Model-agnostic Explanations (LIME) [11], that generate user-tailored explanations, designed to support both technical and non-technical users. This vision extends beyond model performance to address the needs of diverse stakeholders by ensuring explanations are context-aware and comprehensible. Furthermore, the framework assesses the robustness of both model predictions and explanations under adversarial conditions, aiming to mitigate emerging threats that target input data, system processes, and explanation mechanisms. To our knowledge, no prior work unifies malicious URL detection, personalized XAI, and adversarial robustness in a single QR code security pipeline. We also treat explanation integrity as a first-class security asset rather than a usability add-on.

The key contributions of this work are: 1) Introduce a role-specific explanation framework that translates token-level attributions into plain-language risk cues. 2) Combine adversarial training with an ensemble of SHAP and LIME explanations to protect both predictions and explanations. 3) Define a reusable evaluation recipe for fidelity, user clarity, and robustness under attack.

The remainder of this paper is structured as follows: Sect. 2 provides an overview of key concepts and related works, while Sect. 3 describes the vision statement and technical foundation of the proposed framework. Key challenges and open questions are identified in Sect. 4. Finally, this vision paper concludes with a summary and future directions in Sect. 5.

> **Use-case of fake parking QR code:** A scammer stickers a fake QR on a pay-and-display machine. Scanning opens `https://n0rth-tyneside-parking.bank-verify.io`, designed to steal credentials. Our system flags the homograph "n0rth-tyneside-parking" and warns the user in plain language before any redirect.

2 Background and Related Work

2.1 QR Code Security and URL-Based Threats

QR codes, originally developed by Denso Wave [12] for vehicle identification, are now widely used in everyday digital interactions. However, studies indicate that users often assume QR codes are inherently safe, combined with the lack of capable security applications, cybercriminals exploit these factors to fool users and distribute malware and phishing attacks [17]. Therefore, mechanisms for validating embedded URL content are essential prior to user engagement.

Existing solutions have been developed to assess QR code security, incorporating mechanisms through cryptographic schemes [17], QR code image-based detection [2], and URL-based classification [3]. However, these studies face critical constraints since the current cryptographic techniques demand the same application to encode and decode cryptographic QR codes, leading to an additional complexity; image-based approaches are computationally expensive and lack insights into the embedded content; and conventional ML-based URL analysis techniques rely on manual feature extraction, which are less able to cope with evolving cyber threats. Table 1 summarizes the strengths and limitations of these approaches.

2.2 Fine-Tuned Large Language Models for Security Solutions

LLMs are deep neural networks trained on large corpora, capable of understanding and generating language through complex pattern recognition. LLMs have introduced advanced capabilities for cybersecurity, with recent work highlighting their effectiveness in both threat detection and response [1]. Wei et al. [18] present that Transformer-based models outperform traditional feature-engineered models in detecting obfuscated threats in unstructured text, particularly URLs, with high accuracy. However, these LLMs are rarely combined with explanation techniques or defense strategies to assess how and why predictions are made, especially under adversarial conditions involving poisoned datasets or manipulated inputs.

Table 1. Comparison of Existing QR Code and URL-based Security Solutions

Study	Approach	Strength	Limitations
[17]	Cryptographic QR	Strong confidentiality and privacy	Needs compatible apps and exhibits poor scalability
[2]	Image-based detection	Detects visual tampering	Ignores URL semantics and demands intensive computation
[3]	Hybrid ML URL-based classifiers	Learn link-based malicious patterns	Requires manual feature extraction and is vulnerable to emerging attack tactics
This work	**LLM-based model integrated with XAI**	**Deep content understanding and explainable**	Still exploratory with potential high computational cost

2.3 Explainable AI in Cybersecurity

Explainable AI (XAI) improves trust in black-box models by clarifying how predictions are made. XAI methods are broadly divided into intrinsic and post-hoc approaches [20]. Intrinsic models, such as decision trees or logistic regression, are inherently interpretable. Post-hoc methods, like LIME and SHAP, explain predictions made by complex models, such as deep learning. However, XAI methods face limitations, including high computational cost and trade-offs between transparency and privacy. Moreover, attackers can exploit explanations to craft adversarial samples to fool model predictions and interpretations [6]. These vulnerabilities highlight the need for robust and secure integration of XAI methods.

2.4 Adversarial Attacks and Defenses in Explainable AI

XAI methods expose attack surfaces that adversaries can exploit to mislead users or compromise model behavior [15]. Common adversarial techniques include data poisoning, which injects misleading samples into training data [19], and model manipulation, which alters explanation behavior via fine-tuning without reducing accuracy [7]. Backdoor attacks embed hidden triggers during training, activating misclassification or misleading explanations under specific inputs [10]. Also, widely used XAI methods like SHAP and LIME remain vulnerable despite defenses such as robust sampling and regularization [4].

3 Vision and Proposed Framework

3.1 Vision Statement

A comprehensive review reveals critical limitations in mitigating adversarial threats within AI-driven security systems, particularly in detecting malicious URLs embedded in QR codes. Arrieta et al. [5] highlight the need for more advanced XAI tools that can explain AI decisions while maintaining confidentiality and robustness, especially in adversarial settings where sensitive data may be exposed. Recent studies also show that LLMs, such as BERT_based models, effectively capture complex features for malicious URL detection [14]. While

XAI adds transparency to AI systems, tailoring explanations to user needs is crucial to ensure interpretability and trust. Different user roles, including analysts, developers, and non-experts, require explanations suited to their expertise and goals [13]. Literature also identifies that generic explanation formats can reduce user trust and lead to misinterpretation [5]. Hence, personalized explanations aligned with user-specific needs are essential for usability and reliability, especially in threat detection.

Building on these findings, we envision a secure, human-centric AI-based defense system for detecting malicious URLs embedded within QR codes by uniquely integrating fine-tuned LLMs with aggregated ensemble XAI methods. In contrast to existing solutions, this vision emphasizes not just high detection accuracy but also personalized explainability and adversarial robustness.

3.2 Core Components of the Framework

The proposed system, shown in Fig. 1, is a multi-component pipeline with six core phases: QR code decoding, URL tokenization, LLM-based classification, explanation generation, user-specific XAI adaptation, and adversarial robustness evaluation. Each phase contributes to decision-making, transparency, and trust.

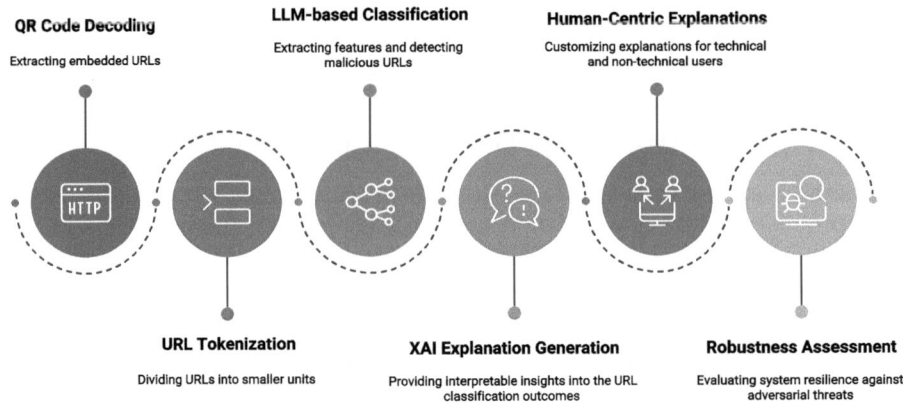

Fig. 1. Core Phases of the Proposed Vision Framework (dotted lines = adversarial channels)

The process begins by decoding the QR code to extract the embedded URL using libraries that convert image-based data into readable text. The decoded URLs are then tokenized into smaller units and passed to an LLM, such as BERT_base model, fine-tuned for malicious URL classification. BERT utilizes an encoder to capture bidirectional contextual information to generate a feature vector from tokenized URLs. On top of the BERT layer, a classification model, such as logistic regression, will be incorporated to predict whether the URL is malicious or benign. Training data will be drawn from publicly available datasets, such as URLhaus, which offer current verified malicious URLs.

To ensure interpretability, the framework adopts post-hoc XAI techniques that estimate which features of the URL contribute to the classification outcome. SHAP and LIME are utilized to analyze the model's behavior through input perturbations and game-theoretic attribution methods, respectively. These explanations are customized for various user roles to enhance clarity and usability across audiences. Human-Computer Interaction (HCI) research and design are becoming critical to provide insights into users' needs and to support the development of more understandable AI systems [8]. Recent literature outlines numerous explanation requirements and notions, yet emphasizes that the optimal structure of effective explanations is still an open problem [16]. Accordingly, we will run semi-structured interviews with analysts, developers, and non-experts. Open coding of the transcripts will map the five dimensions to concrete explanation templates, as illustrated in Fig. 2, implemented as rule-based post-processors over SHAP and LIME outputs.

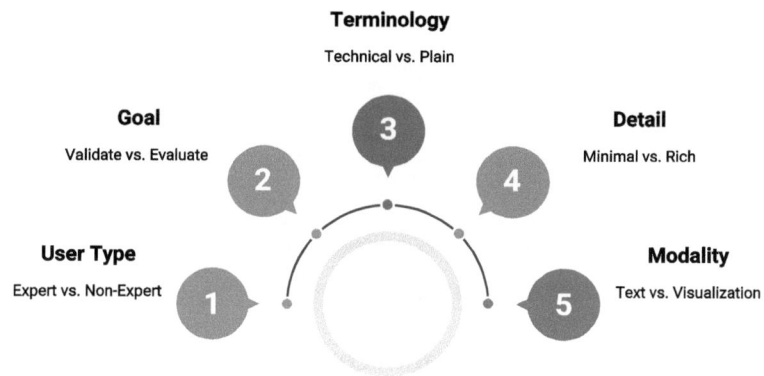

Fig. 2. Key Dimensions Shaping the Structure of User-Specific Explanations

However, revealing feature contributions may expose sensitive model behaviors, making the system and explanations more vulnerable to adversarial threats, as discussed earlier in Sect. 2.4. Addressing this gap, the framework integrates adversarial evaluation to safeguard both predictions and explanations through layered defense strategies. This is accomplished by aggregating explanations produced by both SHAP and LIME utilizing a stacking-based ensemble approach, which in turn reduces variation in prediction errors, making it a more robust interpretation and less susceptible to manipulation targeted at a single XAI method. Future iterations will also test Integrated Gradients and counterfactual methods (e.g., DiCE) and fold them into the ensemble if complementary.

Additionally, adversarial training is incorporated by introducing slightly perturbed input URLs into the training data to improve the model's ability to detect and correctly classify even the manipulated URLs, thus enhancing model robustness. Along with adversarial training, explanation consistency checks will be performed using similarity measures to analyze the stability of explanations under perturbations.

Although still exploratory, this framework sets a concrete research agenda that addresses known limitations in existing work and contributes to the development of explainable, robust, and secure AI-based QR code security. Table 2 summarizes the technical components, tools, and datasets associated with each framework phase. This modular structure is intended to support implementation, testing, and extensibility.

Table 2. Key Components of the Proposed Framework

Component	Proposed Technique	Purpose	Tool/Dataset
QR Code Decoding	Python-based QR libraries	Extract embedded URLs	pyzbar and zxing packages
URL Tokenization	Transformer-based tokenizers	Break URLs into manageable units	HuggingFace AutoTokenizer
URL Classification	Fine-tuned LLM such as BERT-base model	Predict whether URL is malicious or benign	HuggingFace Transformers and URLhaus datasets
Explanation Generation	Post-hoc XAI such as SHAP and LIME	Provide interpretable reasoning for the classification	SHAP and LIME libraries
Personalized Explanations	User modeling by role, including analyst, developer, and non-technical users	Adapt fidelity and format of explanations	Rule-based or user-profile driven
Robustness Assessment	Adversarial input tests and explanation consistency checks	Evaluate resilience of classifier and explainer under attack	TextAttack library and similarity measures

4 Key Challenges and Open Questions

While this vision introduces a promising framework for explainable and secure QR code analysis, various open research and technical challenges remain unexplored in the existing literature. These challenges require further attention to ensure the system is trustworthy, adaptive, and resilient, besides its efficiency in identifying malicious URLs.

One critical challenge is ensuring the explanations, while personalized to different user groups, remain faithful and accurate to the underlying model decisions. Tailoring the explanations enhances user comprehension and trust, but simplified interpretations may omit technical nuances, leading to misunderstanding or misplaced trust. For instance, non-expert users may misinterpret a safe label without understanding conditional factors. A design approach for achieving the proper balance between accessibility and fidelity is an ongoing question.

Another core issue is handling tradeoffs between predictive performance, explanation quality, and system robustness. LLMs and post-hoc XAI methods are computationally intensive, which can affect real-time use cases, including QR scanning on mobile devices. Incorporating explainability layers may also introduce latency or reduce classification accuracy under certain conditions. Ensuring that the system remains lightweight and efficient while maintaining meaningful and trustworthy explanations is a critical area for future research.

Additionally, XAI techniques inherently expose an extra layer in the system that may be targeted by adversaries. By exploring how models think and how a specific outcome is processed, the attacker can construct misleading examples that trick not only the predictions but also the explanation outputs themselves. Consequently, robustness must be ensured at both the classification and explanation levels to mitigate multi-vector threats. These threats remain an open research problem that demands broader investigation from the community.

Table 3 illustrates the key challenges, their impact, and potential corresponding mitigation strategies in the context of malicious QR code defense systems and the use of XAI methods.

Table 3. Summary of Challenges, Impact, and Potential Mitigation Strategies

Challenge	Impact	Potential Mitigation
Tailoring simplified explanations for non-expert users	Distorted understanding, leading to misplaced trust and reduced security if users ignore risks	Design tiered explanation layers or feedback loops to enhance user comprehension
Balancing Performance, Explainability, and Robustness	Increased latency and reduced accuracy in real-time scenarios	Employ lightweight XAI methods and optimize the inference pipeline
Resilience against adversarial manipulation of explanations	Misleading interpretations that may deceive users and evade detection	Incorporate integrity checks for explanations and use adversarial training to defend explanation layers

5 Conclusion and Next Steps

This paper presented a novel and forward-looking direction to securing QR code usage in daily digital activities. As QR codes adapt across various sectors, phishing and malware threats also grow with them. Conventional measures often fail to detect advanced threats and struggle to gain users' trust, a fundamental necessity in developing cybersecurity applications. In this work, we proposed the development of an explainable, human-centric, and secure framework using LLMs to detect malicious URLs embedded in QR codes. By integrating ensemble XAI techniques with adversarial robustness strategies, the introduced framework promises to enhance transparency, adaptability, and user trust in AI-driven security systems. It responds to the growing exploitation of QR codes in cyberattacks and addresses the challenges related to model interpretability and robustness in real-world settings.

In the next phase of this research, we plan to achieve the following key milestones:

1. **Prototype and Baseline:** We will fine-tune a BERT-base model on the URLhaus dataset, aiming for high classification accuracy ($\geq 95\%$ F1), strong explanation fidelity (e.g., delete-insert score ≥ 0.80), and robustness under adversarial conditions ($\leq 5\%$ F1 drop after 100 character-level perturbations). These targets will serve as provisional benchmarks to guide early development and refinement.
2. **Small-Scale User Study:** This will help examine the usability and effectiveness of the proposed system and its tailored explanations, focusing on the perspectives of cybersecurity analysts, system developers, and non-expert end users. Feedback from this user study will inform further enhancements and guide iterative improvements.
3. **Evaluation and Optimization:** This phase will involve assessing the robustness and efficiency of the defense system, including rigorous testing under adversarial conditions. Based on the outcomes, we will perform optimizations to improve model performance, enhance explanation quality, and strengthen resilience against adversarial threats.

At its core, this work is not just about constructing a better detector; it is an invitation to pursue a new direction in AI security: one focused on trust, interpretability, and resilience, and is prepared for deployment in the systems people already rely on. Furthermore, the proposed solution is designed with real-world deployability in mind. The framework can be integrated into mobile-based QR code scanners or enterprise systems to flag malicious URLs at the point of access. Notably, this work also contributes to the broader field of responsible and explainable AI. This vision invites further research into explainable and secure LLM-based cybersecurity systems.

Disclosure of Interests. The authors declare no competing interests relevant to the content of this article.

References

1. Afane, K., Wei, W., Mao, Y., Farooq, J., Chen, J.: Next-generation phishing: how LLM agents empower cyber attackers. In: 2024 IEEE International Conference on Big Data (BigData), pp. 2558–2567 (2024)
2. Alaca, Y., Çelik, Y.: Cyber attack detection with QR code images using lightweight deep learning models. Comput. Secur. **126**, 103065 (2023)
3. Almousa, H., Almarzoqi, A., Alassaf, A., Alrasheed, G., Alsuhibany, S.A.: QR Shield: a dual machine learning approach towards securing QR codes. Int. J. Comput. Dig. Syst. **16**(1), 887–898 (2024)
4. Baniecki, H., Biecek, P.: Adversarial attacks and defenses in explainable artificial intelligence: a survey. Inf. Fusion **107**, 102303 (2024)
5. Barredo Arrieta, A., et al.: Explainable Artificial Intelligence (XAI): concepts, taxonomies, opportunities and challenges toward responsible AI. Inf. Fusion **58**, 82–115 (2020)
6. Charmet, F., et al.: Explainable artificial intelligence for cybersecurity: a literature survey. Ann. Telecommun. **77**(11), 789–812 (2022)

7. Heo, J., Joo, S., Moon, T.: Fooling neural network interpretations via adversarial model manipulation. In: Advances in Neural Information Processing Systems. vol. 32. Curran Associates, Inc. (2019)
8. Liao, Q.V., Varshney, K.R.: Human-centered Explainable AI (XAI): from algorithms to user experiences (2022). arXiv:2110.10790 [cs]
9. Lundberg, S.M., Lee, S.I.: A unified approach to interpreting model predictions. In: Advances in Neural Information Processing Systems, vol. 30. Curran Associates, Inc. (2017)
10. Noppel, M., Peter, L., Wressnegger, C.: Disguising attacks with explanation-aware backdoors. In: 2023 IEEE Symposium on Security and Privacy (SP), pp. 664–681 (2023)
11. Ribeiro, M.T., Singh, S., Guestrin, C.: "Why should i trust you?": explaining the predictions of any classifier. In: Proceedings of the 22nd ACM SIGKDD International Conference on Knowledge Discovery and Data Mining. KDD '16, New York, NY, USA, pp. 1135–1144. Association for Computing Machinery (2016)
12. de Seta, G.: QR code: the global making of an infrastructural gateway. Global Media China **8**(3), 362–380 (2023)
13. Srivastava, G., et al.: XAI for cybersecurity: state of the art, challenges, open issues and future directions (2022)
14. Su, M.Y., Su, K.L.: BERT-based approaches to identifying malicious URLs. Sensors **23**(20), 8499 (2023)
15. Vadillo, J., Santana, R., Lozano, J.A.: Adversarial attacks in explainable machine learning: a survey of threats against models and humans. WIREs Data Min. Knowl. Discovery **15**(1), e1567 (2025)
16. Vilone, G., Longo, L.: Notions of explainability and evaluation approaches for explainable artificial intelligence. Inf. Fusion **76**, 89–106 (2021)
17. Wahsheh, H.A.M., Luccio, F.L.: Security and privacy of QR code applications: a comprehensive study, general guidelines and solutions. Information **11**(4), 217 (2020)
18. Wei, Y., Nakayama, M., Sekiya, Y.: An interpretable fine-tuned BERT approach for phishing URLs detection: a superior alternative to feature engineering. In: 2024 11th International Conference on Social Networks Analysis, Management and Security (SNAMS), pp. 138–145 (2024)
19. Zhang, H., Gao, J., Su, L.: data poisoning attacks against outcome interpretations of predictive models. In: Proceedings of the 27th ACM SIGKDD Conference on Knowledge Discovery & Data Mining. KDD '21, New York, NY, USA, pp. 2165–2173. Association for Computing Machinery (2021)
20. Zhang, Z., Hamadi, H.A., Damiani, E., Yeun, C.Y., Taher, F.: Explainable artificial intelligence applications in cyber security: state-of-the-art in research. IEEE Access **10**, 93104–93139 (2022)

Classification

Exploring Classification with Spectral Transformation

Alexander Stahl

Brandenburg University of Technology, Cottbus, Germany
`stahlale@b-tu.de`

Abstract. Many classification models assign a real-valued score to each object and apply a threshold to determine class membership. While a variety of well-established methods exist for constructing such scores, the use of spectral techniques has received little attention in this context. In this paper, we explore a novel classification approach that treats the label function of binary training data as a signal over the feature space. Using the Discrete Cosine Transform (DCT), we approximate this signal on a sparse grid and reconstruct a smooth decision function whose values are subjected to a fixed threshold. This formulation inherently emphasizes low-frequency components, which promotes smoothness and potentially improves generalization. We discuss the theoretical motivation, implementation challenges, and present experiments that suggest spectral methods may offer an alternative perspective on binary classification.

Keywords: machine learning · classification · discrete cosine transformation · signal approximation · sparse grid · smooth decision boundaries

1 Introduction

Many contemporary classification concepts estimate the likelihood of an object belonging to a class by computing a score or membership probability. These methods differ widely in how they derive such estimates, each offering distinct advantages and trade-offs. However, the use of spectral transforms – such as the Discrete Fourier Transform (DFT) or Discrete Cosine Transform (DCT) – remains largely unexplored in this context, despite their proven utility in signal processing and data compression. This study investigates a classification approach that treats the label function as a signal and uses spectral approximations to produce smooth, generalizable decision boundaries for binary classification.

To motivate the use of spectral methods such as the DCT, it is helpful to consider a well-known application: The JPEG image compression standard, described in [8], relies on the DCT to represent image content in the frequency domain. In this setting, high-frequency components are discarded or heavily quantized, while low-frequency components are preserved. This process achieves compression by retaining only the most perceptually significant information. Drawing a parallel to classification, we propose that a similar trimming of high-frequency components in the spectral representation of the label signal can suppress noisy or overly complex patterns in the training data. By preserving only the smooth, low-frequency structure of the class signal, this approach may improve generalization and provide a natural form of regularization that helps to avoid overfitting.

1.1 Score-Based Binary Classification

Classification is concerned with the assignment of objects to predefined classes, using a model that is trained with labeled data. Here we formulate the general classification problem as the search for a function $K : O \rightarrow C$ such that $K(o) \approx C(o) \; \forall o \in TE$, where $TR \subset O$ and $TE \subset O$ denote the sets of training and test data, respectively, and $C(o)$ the true class label of any object $o \in O$. As we are working with binary classification, the set of possible class labels is $C = \{0,1\}$. We assume that all feature vectors are standardized to the unit hypercube $[0,1]^n$ where n is the number of dimensions. While classification models can differ greatly in terms of their underlying mechanisms, here we focus on methods that approximate a metric score – a real-valued function over the feature space – that can be interpreted as an indicator of class membership. Finally, a threshold may be applied to the score to make the classification decision. This idea is illustrated in Fig. 1. In particular, Fig. 1a depicts a synthetic 2D scenario in which class-1 objects form a circular region in feature space (shaded in blue). Figure 1b shows how such a shape can be approximated by a continuous function over the space, with a decision boundary induced by a threshold value τ (green plane). Objects for which the function output exceeds τ are classified as 1, whereas all others are classified as 0.

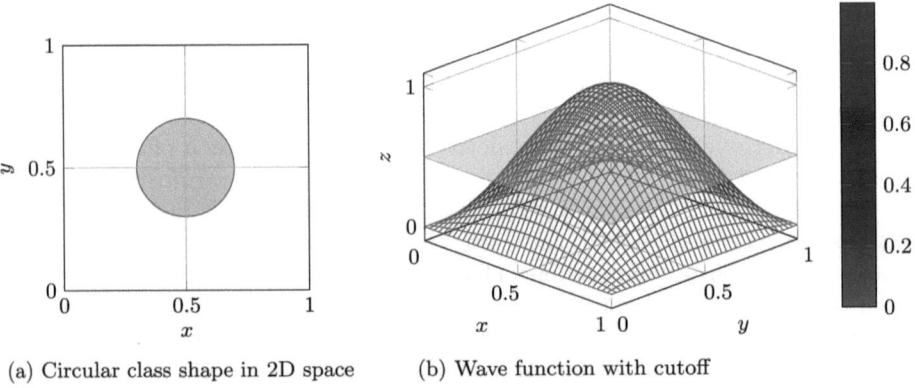

(a) Circular class shape in 2D space   (b) Wave function with cutoff

Fig. 1. Visualization of classification based on score assignment and threshold (Color figure online)

Based on the general concept of assigning a score to each object and applying a threshold to make classification decisions, a wide range of particular classification strategies exists. Each approach comes with its own advantages and limitations, some of which are discussed in the following Sect. 1.2.

1.2 Related Concepts

This section briefly reviews a selection of classification methods that share similarities with our presented approach. The descriptions of logistic regression, naive Bayes, and random forests are based on the presentations in [3,6].

Logistic Regression remains one of the most popular methods for binary classification. For a given object, a metric value is calculated through the usage of linear combination of its weighted feature values: $z = w_1 x_1 + w_2 x_2 + \ldots + w_n x_n$. This value is then subjected to a function such as sigmoid in order to transform it into a probability of class membership. This approach's popularity stems from the simplicity of the calculation and the relative interpretability of the model, which is provided by the weights w_i. These weights can be understood as an evaluation of the importance of each feature for the decision. A limitation of the method lies in its reliance on a linear combination of input features.

Another related approach is the *Naive Bayes* classifier, which applies Bayes' theorem under the assumption that features are conditionally independent given the class label. It computes the posterior probability for each class based on the observed features and assigns the input to the class with the highest probability. Some of the biggest problems of Naive Bayes lie in its assumption of feature independence, its poor calibration of probabilities and the limited expressiveness of its decision boundaries.

Random forests are an ensemble learning method that constructs multiple decision trees on randomly sampled subsets of the training data and features. The prediction is typically made by aggregating the individual tree outputs through majority voting. Like our approach, random forests implicitly partition the input space and combine local decisions into a global classification outcome, where the voting result can be interpreted as an output score. However, they differ in that they produce non-smooth, axis-aligned boundaries and rely heavily on data-driven splits rather than a global approximation of the label function. Their tendency to overfit in high-dimensional settings with sparse data constitutes a major drawback.

Quantum Logic Decision Trees and its derivates function in a similar way, as in that they also calculate a numeric value that is used for the class decision. In contrast to logistic regression, this so-called score is already in the domain of $[0, 1]$ and can immediately be understood as a probability of an object belonging to a class. The score $s(o) = [e]^o$ is the result of the evaluation of an object o with a quantum-inspired logic expression e that is syntactically equivalent to Boolean algebra and therefore offers good interpretability.

$$th_\tau(y) = \begin{cases} 1 & \text{if } P(y = 1 \mid x) \geq \tau \\ 0 & \text{if } P(y = 1 \mid x) < \tau \end{cases} \quad (1)$$

A well-chosen threshold τ is used to decide (see Eq. 1), whether the object has reached a sufficient score to be assigned class membership. QLDT yields interpretable models, but its original formulation suffers from high computational cost during training.

To the best of our knowledge, no classification model integrates the DCT itself. When employed in classification tasks, prior work limits the usage of DCT to preprocessing or feature extraction.

2 Approach Description

For the construction of our classification function, we assume the label $C(o) \in \{0, 1\}$ of the training objects $TR \in [0, 1]^n$ to be a value of signal energy $x(t)$ where x is the label value and t corresponds to the position of o in the n-dimensional feature space. More specifically, we assume that the signal x is 1 at the coordinate of every object with class 1. Likewise, we assume a signal value of 0 at the coordinate of every object of class 0.

2.1 Learning Coefficients via DCT

Given the signal $x(t)$ that represents the class label distribution over a discretized feature space, we transform it from the spatial domain into the frequency domain using the Discrete Cosine Transform (DCT). The formulation and interpretation of the DCT (including the reconstruction of the signal in Sect. 2.5) follow the standard treatment presented in [5]. DCT is preferred over the Discrete Fourier Transform (DFT) in this context for several reasons. First, the DCT operates entirely on real-valued inputs and produces real-valued coefficients, which eliminates the need to work with complex numbers. Otherwise, both computational and conceptual complexity would be increased. Second, the DCT assumes that the input signal is non-periodic and has even symmetry, which aligns better with the nature of label functions in classification tasks: they are typically bounded, defined only over a finite region (the feature space), and do not typically show periodic behavior. In contrast, DFT implicitly assumes periodic repetition of the input signal, which can lead to artificial discontinuities at the boundaries and undesired high-frequency components in the reconstruction.

The DCT of a discrete signal x_n of length N is defined as:

$$X_k = \sum_{n=0}^{N-1} x_n \cdot \cos\left(\frac{\pi}{N}\left(n + \frac{1}{2}\right)k\right), \quad \text{for } k = 0, 1, \ldots, N-1. \qquad (2)$$

The coefficients X_k in Eq. 2 represent the frequency components of the approximated label signal. By discarding high-frequency coefficients and keeping only the lower-frequency terms, we obtain a smoothed approximation that captures the broad structure of class regions while suppressing local noise.

2.2 Sparse Grid Discretization

Both DCT and DFT alike require a uniform sampling of the original signal because they construct the frequency domain from a series of fixed-interval samples. For multidimensional cases, this requirement translates into sampling over a regular grid. This grid has to be explicitly defined in order to allow the application of transformation methods. Since we operate in the standardized domain $[0, 1]^n$, each coordinate of the grid points lies within the interval $[0, 1]$. To construct a grid, we must determine its resolution by specifying the number of points per dimension. We do this by providing a spacing δ that determines the distance between adjacent grid points. For example, setting $\delta = 0.1$ would result in 11 evenly spaced values $(0.0, .0.1, \ldots, 0.9, 1.0)$ along each axis. The choice of δ acts as a hyperparameter of the method: smaller values of δ lead to a finer grid, increasing both computational complexity and model size, but potentially improving accuracy by capturing more detailed spatial structure. For optimal performance, the value of δ should be chosen to reflect the spatial distribution of the training data, ensuring that the grid resolution aligns with the scale at which the data is effectively sampled.

In practice, the feature vectors of observed objects typically do not align exactly with predefined grid points. Therefore, each point must be mapped to its nearest grid location, which can be achieved by rounding the coordinates based on the chosen δ. For this we calculate the grid index g_i for each feature $x_i \in [0, 1]$ as follows:

$$g_i = \min\left(\max\left(\left\lfloor \frac{x_i}{\delta} + 0.5 \right\rfloor, 0\right), \left\lfloor \frac{1}{\delta} \right\rfloor\right) \tag{3}$$

To illustrate this idea, Fig. 2 shows a 2D grid with its grid points depicted in blue. Red points represent the training data and red arrows the alignment to their respectively closest grid point.

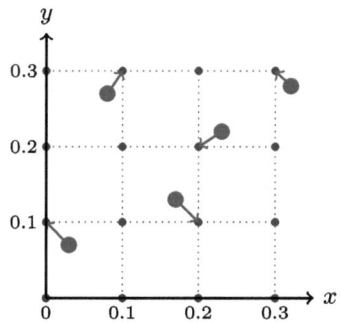

Fig. 2. Visualization of grid alignment

Furthermore, accurate transformations require a dense grid, where every grid point contains a value. However, in most practical settings and especially in tasks involving labeled data, we encounter sparse grids due to the lack of training data for all objects in the domain. To address this, we assume that labeled positive samples (label 1) are sufficiently represented, and we fill in the unlabeled regions by assigning them a default label of 0. This enables us to approximate a dense grid structure and apply transformation techniques despite the possibly incomplete data. Since the zero-labeled objects are now implicitly represented across the grid, we can discard them from the training set.

2.3 Feathering and Interpolation

It is well-known that techniques such as Fourier and cosine transformation struggle with sharp, sudden changes in signal strength. However, as the class labels in our scenario can only have two possible values (0 and 1), this results in a very sharp, abrupt increase around practically every point. Depending on the configuration of the grid, it may also happen that many label-1 points are surrounded by label-0 points, which increases this problem. To counteract this, we introduce feathering to soften the slopes around the affected points and areas. Figure 3 serves as an illustration of this concept, showing the effect of feathering on a 2D grid with a heatmap visualization.

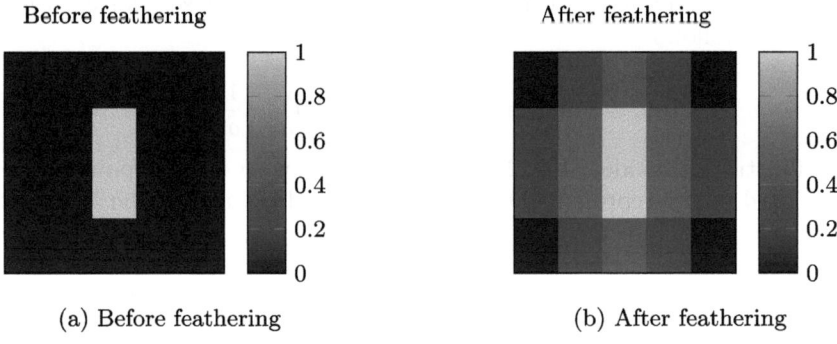

(a) Before feathering (b) After feathering

Fig. 3. Heatmap visualization of a 5 × 5 grid before and after feathering

One of the easiest approaches would be to calculate the average of a point's neighbors' values. This would only affect the direct neighbors of label-1 points. For our use case we desire a more flexible method that allows for configuration of the slope and maximum distance that affected points may have to the respective origin point. Let $f : \mathcal{Z}^n \to \{0, 1\}$ be the original binary signal on the grid and $\mathcal{N}_R(x)$ be the neighbourhood or radius R around a point x. Also, let $K(r)$ be a decay kernel. Then we define the feathered signal:

$$\tilde{f}(\mathbf{x}) = \sum_{\mathbf{r} \in \mathcal{N}_R(\mathbf{0})} f(\mathbf{x} - \mathbf{r}) \cdot K_\sigma(\mathbf{r}) \qquad (4)$$

Another alternative is to use the concept of k-nearest-neighbors (kNN) to determine the most appropriate label for each point based on the majority label among its neighbors:

$$\hat{y} = \arg\max_{c \in \mathcal{C}} \sum_{i \in \mathcal{N}_k(x)} 1(y_i = c) \qquad (5)$$

where $1(\cdot)$ denotes the indicator function and $\mathcal{N}_k(x)$ denotes the set of the k nearest neighbors of the query point x in the training set. This method does not produce continuous values in the interval $(0,1)$, and therefore cannot be considered a form of feathering. However, it may reduce the number of empty regions between falsely isolated points.

2.4 Determination of Output Threshold

Due to the differences in predicted signal values for label-1 objects, it is not suitable to use a fixed threshold of e.g. $\tau = 0.5$. Instead, the training set TR is used to determine a well-working value for τ. Here we follow the approach described in [7], which we will only mention shortly for the purpose of completeness. Assuming that the evaluation result of an object is its estimated energy as predicted via the DCT coefficients, let $\min_O$ be the smallest evaluation result of the 1-objects and $\max_Z$ the highest result of the label-0 objects from TR. Then τ is calculated as shown in Eq. 4. A more detailed discussion is provided in [7].

$$\tau = \begin{cases} \dfrac{\max_Z + \min_O}{2} & \text{if } \max_Z < \min_O, \\ \arg\max_{\tau_o \in [\min_O, \max_Z]} (\text{accuracy}(e, \tau_o, TR)) & \text{otherwise.} \end{cases} \qquad (6)$$

2.5 Classification of Objects

The trained classification model consists of the determined set of DCT coefficients and the output threshold τ. To classify a new object o, the coefficients are used to reconstruct the signal at the location corresponding to o in the discretized input grid. This reconstructed value represents the signal strength or score at that point and is computed as:

$$s(o) = \sum_{k_1=0}^{N_1-1} \cdots \sum_{k_n=0}^{N_n-1} X_{k_1,\ldots,k_n} \prod_{j=1}^{n} \cos\left(\frac{\pi}{N_j}\left(t_j + \frac{1}{2}\right)k_j\right) \qquad (7)$$

where $X_{k_1,\ldots,k_n}$ are the learned DCT coefficients, N_j is the number of grid points in dimension j and t_j is the (scaled and discretized) coordinate of object o in dimension j. The resulting signal score $s(o)$ is then compared to a threshold, as defined in Eq. 1, to assign a class label.

3 Experimental Evaluation

3.1 Implementation

For testing purposes, a prototype of our approach was implemented in Python 3.10. A custom DCT implementation had to be used due to the limitations of the typically used libaries for DCT such as numpy.fft or scipy.fft. This is mainly related to the implementation of the sparse grid. The aforementioned libraries are unable to operate on sparse grids and therefore require an instantiation of the full dense grid in memory. Since the required memory for such a grid of float64 values would be $m^n \times 8$ bytes, (with n dimensions and m grid values per dimension), the amount of storage space required quickly exceeds any size available in practice.

To circumvent this, our implementation assigns a default value (typically 0) to any grid point not explicitly stored. Since the output of the transform is also represented as a dense grid of values, we impose a maximum size constraint on the output to manage memory and computation requirements. Values corresponding to higher-frequency components beyond this limit are imediately omitted in the calculation process.

3.2 Experiment Setup

We evaluate the presented classification method by applying it to various datasets and evaluating its performance, using various metrics. Experiments were performed on a regular laptop (Intel Core i7-7500U CPU, 16 GB RAM, Ubuntu 22.04.5 LTS). and were set up using Python and the scikit-learn library [4].

Our evaluation is done in comparison with two established baseline models: k-Nearest Neighbors (KNN) and Gaussian Naive Bayes (GNB). The experiments are conducted across four binary classification tasks: three derived from standard benchmark datasets – Breast Cancer [9], Iris [2], and Wine [1] – and a synthetic dataset generated with scikit-learn's make_classification function. For the Iris and Wine datasets, binary labels are constructed by treating class 0 as the positive class and merging all remaining classes into the negative class. All datasets are normalized to the $[0, 1]$ interval.

To limit the computational cost, the dimensionality of the data was reduced to 5 before applying it to our classification model. This was done using scikit-learn's SelectKBest class. The hyperparameter δ is optimized using grid search over each cross-validation fold. The baseline models are trained using their default parameter configurations. Following training, the models are evaluated on the test set using standard metrics: accuracy, precision, recall, F1 score, and ROC AUC. Additionally, we report the standard deviation to assess the consistency of model performance. We evaluate the following metrics for each dataset: classification accuracy, precision and recall, F1 score, AUC and generalization error. The metrics are compared to baseline values from kNN and DT. All experiments were done with k-fold cross validation, using $k = 5$.

Table 1 presents the results of our experiments, with "STC" denoting the classifier introduced in this paper. The experimental results demonstrate that STC

Table 1. Results of Experimental Evaluation

Dataset	Classifier	CV σ	Acc.	Prec.	Rec.	F1 Score	ROC AUC
Breast Cancer	KNN	0.0164	0.9473	0.9452	0.9718	0.9583	0.9872
	Naive Bayes	0.0256	0.9649	0.9718	0.9718	0.9718	0.9954
	STC	0.2303	0.9122	0.9420	0.9154	0.9285	0.887651
Iris	KNN	0.0	1.0	1.0	1.0	1.0	1.0
	Naive Bayes	0.0	1.0	1.0	1.0	1.0	1.0
	STC	0.0988	1.0	1.0	1.0	1.0	1.0
Wine	KNN	0.0173	1.0	1.0	1.0	1.0	1.0
	Naive Bayes	0.014	1.0	1.0	1.0	1.0	1.0
	STC	0.0267	0.8333	1.0		0.5714	0.7272	0.8392
Synthetic	KNN	0.0142	0.72	0.6538	0.7727	0.7083	0.800
	Naive Bayes	0.0251	0.735	0.6667	0.7954	0.7253	0.8095
	STC	0.0818	0.67	0.5846	0.8636	0.6972	0.6558

achieves competitive performance across a variety of datasets. On well-structured datasets such as Iris, all classifiers, including STC, reached perfect scores. For the Breast Cancer dataset, STC performed slightly below the baselines in terms of accuracy and AUC but maintained a strong F1 score of 0.93, which indicates reliable overall prediction quality. On the synthetic dataset, STC achieved the highest recall (0.86), suggesting that it is effective at detecting positive cases, albeit with some loss in precision. The Wine dataset proved more difficult for STC, likely due to its finer-grained class structure. It is worth noting that, due to computational limitations, feathering was not applied in these experiments. Since this step is designed to enhance smoothness and generalization, future evaluations incorporating feathering may further improve performance. Overall, these results support the potential of spectral classification as an alternative to traditional methods.

4 Conclusion

This paper introduced a spectral classification approach that treats the label function of a binary classification problem as a signal over the input space. By applying the Discrete Cosine Transform (DCT) to this label signal, we construct a smooth, global approximation of class membership that supports score-based classification via thresholding.

Aside from various score-based classifiers, a key conceptual contrast can be drawn to k-nearest neighbors (kNN), where classification is based on local label comparisons in the vicinity of a point. In our method, the same training data is instead interpreted in the spectral domain. Rather than memorizing label information locally, we compute a global approximation of the label function. This means that a change in any individual point affects the signal reconstruction across the entire input space. The approach inherently favors smooth decision boundaries and can be viewed as learning a compressed, low-frequency representation of class structure.

Among its advantages, this method is robust to label noise, encourages generalization by suppressing high-frequency components, and avoids storing individual data points. It may be especially well suited for structured domains where class membership varies smoothly across the feature space – a property often found in real-world tasks. However, its reliance on a grid structure presents challenges in high-dimensional settings, where memory requirements grow rapidly. This issue can be mitigated by using sparse grids with a default value for unobserved points, although this makes the implementation less compatible with standard libraries and complicates training.

Whether the proposed approach consistently improves generalization will require further empirical study. Nonetheless, we believe it offers a promising and underexplored perspective on classification. Preliminary results suggest that spectral approximation of label signals may serve as a viable alternative to more conventional methods.

Disclosure of Interests. The authors have no competing interests to declare that are relevant to the content of this article.

References

1. Aeberhard, S., Coomans, D., de Vel, O.: Wine data set. UCI Machine Learning Repository (1992). https://archive.ics.uci.edu/ml/datasets/wine
2. Fisher, R.A.: Iris. UCI Machine Learning Repository (1936). https://doi.org/10.24432/C56C76
3. Han, J., Kamber, M., Pei, J.: Data Mining: Concepts and Techniques, 3rd edn. Morgan Kaufmann, Boston (2011)
4. Pedregosa, F., et al.: Scikit-learn: machine learning in python. J. Mach. Learn. Res. **12**, 2825–2830 (2011)
5. Rao, K.R., Yip, P.: Discrete Cosine Transform: Algorithms, Advantages, Applications. Academic Press, Boston (1990)
6. Russell, S., Norvig, P.: Artificial Intelligence: A Modern Approach, Global Pearson Education, London (2021)
7. Schmitt, I.: QLC: a quantum-logic-inspired classifier. In: Proceedings of the 26th International Database Engineered Applications Symposium. IDEAS '22, New York, pp. 120–127. Association for Computing Machinery (2022). https://doi.org/10.1145/3548785.3548790
8. Wallace, G.: The jpeg still picture compression standard. IEEE Trans. Consum. Electron. **38**(1), xviii–xxxiv (1992). https://doi.org/10.1109/30.125072
9. Wolberg, W., Mangasarian, O., Street, N., Street, W.: Breast Cancer Wisconsin (Diagnostic). UCI Machine Learning Repository (1993). https://doi.org/10.24432/C5DW2B

Optimizing Classification Accuracy with Simulated Annealing in k-Anonymity

Despina Tawadros[1(✉)], Wenhui Yang[1], Lena Wiese[1,2], and Volker Meyer[2]

[1] Fraunhofer Institute for Toxicology and Experimental Medicine,
Nikolai-Fuchs-Str. 1, 30625 Hannover, Germany
{despina.tawadros,wenhui.yang}@item.fraunhofer.de,
lwiese@cs.uni-frankfurt.de

[2] Institute of Computer Science, Goethe University Frankfurt, Robert-Mayer-Str. 10, 60325 Frankfurt am Main, Germany
v.meyer@em.uni-frankfurt.de

Abstract. In the era of extensive data collection, achieving a balance between individual privacy protection and the preservation of data utility is critical. This paper introduces a novel k-anonymization approach that integrates simulated annealing with generalization hierarchies and suppression constraints to optimize classification accuracy on anonymized datasets. Unlike traditional greedy algorithms, our method probabilistically navigates the anonymization solution space. We validate our approach through extensive experiments on two real-world datasets, Adult and MIMIC-III, comparing against the state-of-the-art ARX framework. Our method improves AUC-ROC scores by up to 3.3% over ARX, and successfully generates feasible anonymizations even under stringent privacy requirements where ARX fails – demonstrating robustness and effectiveness of our simulated annealing-based anonymization strategy.

Keywords: k-anonymity · Privacy-preserving · Classification analysis

1 Introduction

In the digital age, web users, during their interactions with various applications and services, end up providing a substantial amount of personal information. This data often contains sensitive elements such as addresses, ages, medical histories, and other private attributes. Analysis of such data have become a cornerstone for numerous companies and organizations. Organizations and researchers increasingly rely on extensive datasets, sharing and exploiting them through advanced data mining techniques to pursue a broad spectrum of objectives.

Problem Statement and Standard Practices. In the process of personal data collection, a standard practice involves the removal of unique identifiers associated with each user to obscure their identity. This method, however, may not adequately address privacy concerns, as simply eliminating unique identifiers does

---

D. Tawadros and W. Yang—Contributed equally to this work.

© The Author(s), under exclusive license to Springer Nature Switzerland AG 2026
G. Bergami et al. (Eds.): IDEAS 2025, LNCS 15928, pp. 71–84, 2026.
https://doi.org/10.1007/978-3-032-06744-9_6

not guarantee the complete anonymity of the data [24]. Anonymized datasets can be re-identified by cross-referencing them with information obtained from other sources [16]. Furthermore, the advent of advanced predictive models has introduced a new level of complexity. The application of machine learning algorithms to anonymized data, despite its altered state, can inadvertently reveal sensitive information about individuals within the dataset. Anonymization techniques involve altering or masking sensitive information within datasets. To this end, the field has developed various privacy-preserving models, such as k-anonymity [21], l-diversity [14], t-closeness [13], δ-differential privacy [6]. k-anonymity's simplicity and effectiveness keep it a frequent choice for organizations and researchers dealing with sensitive data. In a k-anonymized dataset, each record is indistinguishable from at least $k-1$ other records with respect to certain identifying attributes. This is achieved by modifying or generalizing quasi-identifiers (like age, zip code, etc.) in the dataset. However, a major issue with data anonymization is its potential impact on the utility of data. As the extent of anonymization increases, it can impact the results of data mining and classification [18].

Research Gap and Contribution. Existing anonymization approaches, including optimization-based methods, have primarily focused on minimizing information loss or achieving theoretical privacy guarantees, often at the expense of practical classification utility. Moreover, few methods explicitly incorporate suppression constraints into the optimization process, which are critical in many real-world applications where aggressive suppression is not feasible. This gap motivates the development of our classification-driven, suppression-aware anonymization strategy. This paper introduces a k-anonymization algorithm based on the simulated annealing algorithm. Our goal is to find a balance between protecting privacy and ensuring the practicality of the data processed by classifiers.

Outline. The remaining sections of this paper are structured as follows. Section 2 describes the background and theoretical framework. Section 3 presents the related work. Section 4 presents our approach. Section 5 provides a comprehensive evaluation of our method, using the real datasets. Finally, Sect. 6 concludes the paper, summarizing our contributions and presenting future work.

2 Background and Preliminary Definitions

We introduce concepts related to k-anonymity as well as optimization metrics applied within the context of k-anonymity algorithms.

2.1 k-Anonymity

Given a dataset D and a set of quasi-identifier attributes $A = \{a_1, a_2, \ldots, a_n\}$:

- A dataset D satisfies **k-anonymity** with respect to a set of quasi-identifier attributes A if for every record r in D, there exist at least $k-1$ other records in D such that all k records are indistinguishable based on A. Formally:
  $\forall r \in D, \quad |\{r' \in D \mid r \neq r', r'[A] = r[A]\}| \geq k-1$

Table 1. Example of original and anonymized data after suppression

Name	Age	Gender	Diseases	Name	Age	Gender	Diseases
Jane	20	female	Drug abuse	****	20	female	Drug abuse
Tom	35	male	Cancer	****	35	male	Cancer
Jack	51	male	Hypertension	****	51	male	Hypertension
Ella	47	female	Diabetes	****	47	female	Diabetes

- A **k-equivalence class** is a subset of D where all records in the subset are indistinguishable from each other based on A, and the size of this subset is at least k. For a given subset $E \subseteq D$, E is a k-equivalence class if: (1) $\forall r, r' \in E$, $r[A] = r'[A]$ and (2) $|E| \geq k$

2.2 Generalization and Suppression Strategy

In our approach, generalization and suppression are treated as dynamic optimization components, rather than static anonymization operations. Instead of predefining fixed levels of abstraction, we leverage simulated annealing to actively explore and select combinations of generalization levels.

Generalization Mechanism: Given a dataset D with a set of quasi-identifiers $A = \{a_1, a_2, \ldots, a_n\}$, each a_i is associated with a Domain Generalization Hierarchy (DGH), which organizes possible generalization levels hierarchically. Formally, generalization is modeled as a mapping: $f_i : \mathcal{D}_{a_i} \to \mathcal{D}_{a_i}^{(g)}$ where $\mathcal{D}_{a_i}$ denotes the original domain of a_i, and $\mathcal{D}_{a_i}^{(g)}$ denotes its generalized domain at level g. A complete generalization solution S is represented as a vector of generalization levels across all quasi-identifiers: $S = (g_1, g_2, \ldots, g_n)$ where g_i is the selected generalization level for a_i. The search space consists of all feasible combinations of generalization levels respecting k-anonymity and suppression limits.

Suppression Strategy: Suppression is incorporated as a controlled fallback mechanism. If, after applying a candidate generalization solution S, some records cannot be assigned to k-equivalence classes without violating k-anonymity, those records may be suppressed. However, suppression is strictly constrained by a user-defined maximum threshold $Supp_{max}$: $Supp(S) \leq Supp_{max}$ where $Supp(S)$ denotes the fraction of suppressed records under solution S. Any candidate solution exceeding the suppression limit is automatically discarded during search.

Solution Space and Search: The set of all feasible generalization vectors S forms a generalization lattice, where moving between nodes corresponds to modifying the generalization level of one or more attributes. Rather than exhaustively traversing the lattice, we employ simulated annealing to efficiently navigate the space, probabilistically accepting both improving and, under certain conditions, non-improving solutions, to escape local optima.

Objective: Our optimization goal is to maximize classification performance (measured by AUC-ROC) while satisfying k-anonymity and suppression constraints: $\max_S$ AUC-ROC(S) subject to k-anonymity and $Supp(S) \leq Supp_{max}$

This dynamic, integrated treatment of generalization and suppression allows our method to achieve high utility anonymized datasets without compromising privacy guarantees. The construction of a generalization hierarchy facilitates the organization of these anonymization levels in a structured manner, enabling the navigation of the trade-off between preserving data usability and safeguarding individual identities. Figure 1 shows generalization hierarchies for the quasi-identifiers in the example dataset. The three quasi-identifiers (age, gender, diseases) from Table 1 are generalized from specific values to more abstract ranges. Suppression involves concealing information by eliminating details (columns or tuples), which entails completely removing records from the final output. Table 1 shows the anonymized data after suppression is applied.

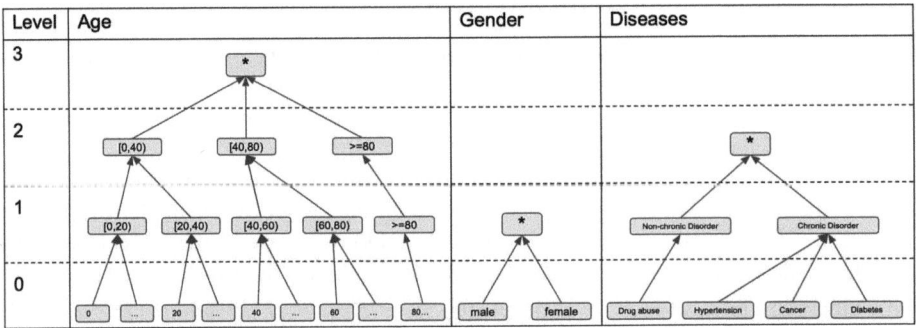

Fig. 1. Value generalization hierarchies for three common quasi-identifiers

2.3 Optimization Metrics

As mentioned earlier, k-anonymity algorithms aim to find the global optimum. Here, 'optimal' refers to the solution that results in the least information loss according to a specific metric. Thus, the transformation of the dataset can be described as optimizing the metric while satisfying the constraints of k-anonymity. [9] points out that the optimization of metrics (i.e., the selection of the objective function) should be determined based on the intended use of the k-anonymized data. One of the early metrics commonly employed is the Precision Metric [20], which quantifies information loss exclusively through its generalization levels, independent of the actual input dataset. Another frequently utilized metric is the Discernability Metric [2]. This metric imposes a penalty on each tuple based on the number of tuples in the transformed dataset that are indistinguishable from it. [9] proposed another optimization metric, classification metric, based on predictive modeling use. This metric does not penalize an unsuppressed tuple if it is part of the majority class within its respective equivalence class; however, it assigns a penalty value of 1 to all other tuples. As shown in Eq. (2), further discussion can be found in the following section.

In this study, we employ the AUCROC score as the optimization objective. The AUCROC is the area under the ROC curve, which graphically represents the performance of a classification model at all classification thresholds. This area can be interpreted as the probability that the model correctly distinguishes between a randomly selected positive example and a randomly selected negative example. AUCROC is not affected by changes in the distribution of the test data. It is especially useful for imbalanced datasets, a common scenario in fields like fraud detection or medical diagnosis, where one class is much rarer than the other. Comparing the scores before and after dataset anonymization directly reflects the impact of various anonymization processes on model accuracy and provides an assessment of the quality of the anonymized datasets.

3 Related Work

The k-anonymity privacy model, introduced by Sweeney [21], remains a foundational technique for protecting personal information. Numerous algorithms have been proposed for achieving k-anonymity, which broadly fall into two categories: local recoding and full-domain generalization [1]. A structured comparison of representative methods and their evaluation metrics is summarized in Table 2.

Local recoding methods apply different generalization rules to the same attribute values across different records, offering flexibility but often at a high computational cost. Clustering-based local recoding techniques, such as greedy clustering [22] and African vultures optimization [19], partition datasets into equivalence classes based on quasi-identifiers. Additionally, metaheuristic approaches including genetic algorithms [8] have been explored to optimize anonymization by balancing privacy and utility. While these methods are effective at achieving anonymity, they often struggle with scalability to high-dimensional datasets and may result in inconsistent generalization patterns, complicating downstream classification tasks. In contrast, full-domain generalization methods apply uniform generalization rules across entire attributes. Notable examples include Incognito [12], OLA (Optimal Lattice Anonymization) [7], and the Flash algorithm used in the ARX anonymization tool [11]. These methods leverage generalization lattices and monotonicity properties to efficiently prune the search space. However, their optimization objectives typically focus on minimizing information loss or discernability penalties without explicitly considering classification utility, and they often permit high suppression rates without explicit constraints [17]. Simulated annealing has also been adapted for k-anonymity, as first proposed by Winkler [23]. The algorithm iterates over various fields in the dataset, evaluating potential changes to each record's value-states based on an objective function and a corresponding penalty function [9] defined as follows:

$$f(\text{value-state}) = \frac{\sum_{\text{all rows}} \text{Penalty}(r)}{N} \quad (1)$$

$$\text{Penalty}(r) = \begin{cases} 1 & \text{if } r \text{ is suppressed} \\ 1 & \text{if class}(r) \neq \text{majority}(G(r)) \\ 0 & \text{otherwise} \end{cases} \quad (2)$$

Here, f(value-state) quantifies the average penalty across all records. A row r is penalized if it is suppressed or if its class label class(r) does not match the majority class label majority($G(r)$) of its equivalence class $G(r)$. While this approach attempts to balance suppression and class consistency, it operates at the individual cell level, leading to substantial computational overhead and lacking explicit mechanisms for controlling maximum suppression. Furthermore, the optimization focuses primarily on penalty minimization without aligning directly to classification performance metrics. Recent work has explored multi-objective optimization strategies, such as utilizing relaxed functional dependencies and Pareto front exploration [4]. Although these methods offer improved flexibility, selecting optimal solutions from large Pareto sets remains challenging for data owners, and classification utility is not directly optimized.

Table 2. Summary of anonymization algorithms

Lit.	Privacy	Technique	Methods	Utility Metrics
[22]	k-anon.	local recoding general.	greedy clustering	Info loss, F-measure
[19]	k-anon., t-close.	local recoding general.	African vultures optimization clustering	Clustering acc., diversity, anonymity
[11]	k-anon.	suppression, full-domain general.	greedy depth-first search, binary search	Discernability, execution time
[23]	k-anon.	local recoding general.	simulated annealing	Classification metric
[4]	k-anon.	full-domain general.	RFDs, Pareto principle	Info gain, accuracy
Ours	k-anon.	suppression, full-domain general.	simulated annealing	AUC-ROC score

While simulated annealing was applied to k-anonymity [23], earlier efforts primarily operated at the cell level, leading to significant computational overhead and lacking explicit suppression control mechanisms. Furthermore, optimization methods such as Pareto front exploration [4] and greedy lattice-based techniques [11] have generally prioritized minimizing general information loss without direct alignment to predictive model performance. Differential Privacy (DP) [6] has also gained attention as a formal privacy framework, but it typically relies on noise injection and randomized algorithms, which can degrade utility and complicate interpretability. In contrast, our method introduces three key advancements: (1) direct optimization of classification utility through the AUC-ROC score, (2) enforcement of suppression limits during anonymization to balance utility and privacy constraints, and (3) a temperature-guided simulated annealing framework exploring generalization lattices efficiently while avoiding local optima.

4 Methodology

4.1 Problem Formulation and Proposed Approach

K-anonymity is highly regarded for its simplicity and effectiveness in protecting personal identities; however, it presents a significant challenge in balancing privacy protection and data utility. The utility of data is a subjective metric, dependent on the specific goals of the data user, and is crucial in evaluating the feasibility of anonymized datasets. In our work, classification accuracy serves as the primary indicator of data utility, reflecting the model's ability to identify patterns and derive meaningful insights from anonymized data. Increasing the k value enhances privacy by obscuring individual identifiability but often leads to a decrease in utility due to the need for stricter generalization and suppression. Such distortion can introduce bias into machine learning models, thus affecting prediction reliability. This is attributed to (1) a decrease in granularity due to generalization, which combines individuals with diverse characteristics into homogeneous groups, thereby concealing subtle patterns; (2) the suppression of excessive data points, limiting the information available for analysis and reducing utility. The complexity of achieving optimal generalization grows rapidly with the number of quasi-identifier attributes and their potential generalization levels, categorizing it as an NP-hard problem [15]. We propose using a simulated annealing-based metaheuristic to efficiently explore the solution space for near-optimal anonymization configurations within feasible computational timescales.

Algorithm 1 presents the pseudocode of our method consisting of:

1. *Data Preparation and Initialization:* We begin by preprocessing the dataset, which involves class rebalancing through imputation, removal of outliers, and the excision of direct identifiers such as IDs and names. Selected Quasi-Identifiers (QIDs) are associated with a Domain Generalization Hierarchy (DGH). We also specify constraints including the maximal suppression percentage ($Supp_{max}$) and the k parameter. The algorithm is initialized with a random solution S_0, representing random generalization levels for each QID. Hyperparameters include the initial temperature T_0, the cooling rate α, and the minimum temperature $T_{\min}$.
2. *Temperature Management:* The algorithm begins with temperature T_0 and updates it exponentially at each iteration i: $T_i = T_0 \cdot \alpha^i$ where $0 < \alpha < 1$ controls the rate of cooling. A high initial temperature ($T_0 = 1000.0$) was selected to enhance exploration in the early search stages.
3. *Termination Conditions:* The process terminates when either the maximum number of iterations is reached without improvement or the temperature falls below the threshold $T_{\min}$. Each temperature level was maintained for 10 iterations to allow sufficient local search; termination was enforced either for temperature below $T_{\min} = 0.001$ or no improvement over 80 evaluations.
4. *Exploration and Solution Evaluation:* During each iteration, the algorithm explores the neighborhood of the current solution by randomly selecting a QID and modifying its generalization level. Formally, given a set of QIDs $A = \{a_1, a_2, ..., a_n\}$, an attribute a_i is selected at random ($1 \leq i \leq n$), and a generalization level $g_j \in G(a_i)$ is randomly chosen ($1 \leq j \leq m$).

The modified dataset is evaluated by partitioning it into k-equivalence classes, calculating suppression levels, and measuring classification performance using the AUC-ROC score. The objective function is defined as: $F(S) = $ AUC-ROC(S) as described in Sect. 5.2.

5. *Metropolis Criterion:* Neighboring solutions are accepted based on the Metropolis criterion: $p = e^{-\Delta F/T_i}$ where ΔF is the change in objective function value. This probabilistic acceptance mechanism allows for occasional acceptance of inferior solutions to escape local optima.
6. *Iteration and Convergence:* As the temperature lowers, the acceptance probability for worse solutions decreases, shifting the search strategy from exploration to exploitation and progressively refining toward the optimal solution.
7. *Output:* Upon termination, the best-discovered solution is output balancing k-anonymity constraint and maximum suppression limit.

Algorithm 1. Optimizing Accuracy with Simulated Annealing in k-anonymity

---
1: **Input:** Dataset D, Quasi-identifiers A, Set of generalization levels G, k, Initial temperature T_0, Cooling rate α, Maximum suppression $Supp_{max}$
2: **Output:** Near-optimal solution S maximizing classification score
3: Initialize S_0 randomly
4: Initialize $T_i \leftarrow T_0$
5: **while** Not converged **do**
6:     Randomly perturb S_0 to generate a neighboring solution S_n
7:     Partition D according to S_n into k-equivalence classes
8:     Calculate suppression percentage $Supp(S_n)$
9:     **if** $Supp(S_n) \leq Supp_{max}$ **then**
10:         Evaluate classification scores $F(S_n)$ and $F(S_0)$
11:         Calculate $\Delta F = F(S_n) - F(S_0)$
12:         **if** $\Delta F \geq 0$ **then**
13:             Accept S_n as the new current solution
14:         **else**
15:             Accept S_n with probability $p = e^{-\Delta F/T_i}$
16:         **end if**
17:     **else**
18:         Discard S_n and generate a new neighbor
19:     **end if**
20:     Update temperature: $T_i \leftarrow \alpha \cdot T_i$
21: **end while**
22: **Return:** Best solution S

More formally, let S denote a candidate anonymization solution, represented by a vector of generalization levels for each quasi-identifier attribute. The search space $\mathcal{S}$ consists of all feasible solutions satisfying k-anonymity and suppression constraints, formally: $\mathcal{S} = \{S \mid S \text{ satisfies } k\text{-anonymity and } Supp(S) \leq Supp_{max}\}$. The optimization objective is to find: $\max_{S \in \mathcal{S}} F(S)$ where $F(S)$ represents the AUC-ROC classification score obtained on the dataset anonymized according to S. The solution space $\mathcal{S}$ is explored via random perturbations of generalization levels, with acceptance of new candidate solutions governed by the Metropolis criterion: $p = e^{-\Delta F/T_i}$, where ΔF is the change in classification score and T_i is the current temperature in the annealing schedule.

4.2 Security Analysis

Dividing the dataset into k-equivalence classes ensures each class contains at least k indistinguishable records. By integrating the simulated annealing search for optimal generalization, the anonymized data exhibits additional randomness, increasing difficulty for attackers attempting linkage or inference-based attacks. Optimizing the distribution of sensitive attributes within equivalence classes further reduces vulnerabilities to statistical attacks. Thus, our anonymization framework defends against multiple threat models:

- **Linkage Attacks**: Ensuring each equivalence class contains at least k indistinguishable records, direct linkage using quasi-identifiers is mitigated.
- **Background Knowledge Attacks**: Randomized exploration of generalization hierarchies reduces deterministic patterns, making it harder for adversaries with external knowledge to accurately infer sensitive information.
- **Inference Attacks**: By balancing suppression and generalization, and by optimizing sensitive attribute distribution within equivalence classes, the method minimizes risks of statistical inference-based re-identification.

The use of randomized optimization via simulated annealing inherently introduces uncertainty into the data transformation process, further complicating attackers' ability to predict the original data structure.

4.3 Complexity Analysis

Computational complexity is determined by the size of the dataset, the number of quasi-identifiers, and the structure of the generalization hierarchies. Each iteration primarily consists of two operations: partitioning the dataset into k-equivalence classes and evaluating the classifier performance. For a dataset with n records and m quasi-identifiers, the per-iteration complexity is approximately $\mathcal{O}(n \times m)$. Simulated annealing's convergence properties are theoretically guaranteed: under a sufficiently slow (logarithmic) cooling schedule, global optimality is achievable. Practically, exponential cooling schedules are used to ensure computational feasibility, resulting in high-quality near-optimal solutions. The Metropolis acceptance criterion allows probabilistic acceptance of inferior solutions at higher temperatures, thereby enabling broader exploration and reducing the risk of getting trapped in local optima—unlike purely greedy methods such as FLASH [11]. Lightweight evaluation techniques (subsampling/incremental model updating) can reduce the computational burden per iteration.

5 Evaluation

5.1 Datasets and Settings

In this evaluation, we compare the performance of our method with the ARX anonymization tool [11] across two real-world datasets focusing on classification tasks. All experiments were conducted on Google Colab using its default

CPU environment (Intel Xeon CPU with 2 virtual CPUs and 13 GB RAM). We employed the Adult dataset and the MIMIC-III dataset to assess the effectiveness of our method. The Adult dataset [3], extracted by Ronny Kohavi and Barry Becker from the 1994 Census database, is a widely used public benchmark for machine learning tasks. After preprocessing, it contains eight attributes identified as quasi-identifiers, including age, work class, education level, marital status, occupation, race, sex, and native country. The target attribute is whether the annual income exceeds $50,000. The MIMIC-III (Medical Information Mart for Intensive Care III) dataset [10] is a large, freely available clinical database comprising de-identified health-related data for over forty thousand patients admitted to critical care units at the Beth Israel Deaconess Medical Center between 2001 and 2012. We selected nine attributes as quasi-identifiers, with the target attribute being death_365_days, indicating whether a patient passed away within one year. Generalization hierarchies were manually constructed based on domain expertise. For instance, age is generalized into decade ranges, and geographic regions follow a nested structure. Both datasets were tested without anonymization to obtain baseline scores; see Table 3.

Table 3. Dataset characteristics and baseline classification scores and effect of cooling rate α on AUC-ROC performance on Adult dataset

Dataset	# QIDs	Records	Baseline AUC-ROC
Adult	8	30,162	0.760
MIMIC-III	9	2,655	0.782

Cooling Rate α	AUC-ROC Score
0.85	0.732
0.90	0.745
0.95	0.748

5.2 Classification Accuracy and Sensitivity Analysis

We utilized CatBoost [5] for binary classification tasks on the anonymized datasets, addressing two main predictive challenges: predicting whether an individual's annual income surpasses $50,000, and forecasting patient survival within one year. The AUC-ROC (Area Under the Receiver Operating Characteristic curve) score was employed as the primary metric to assess classification performance due to its robustness in imbalanced datasets such as MIMIC-III. Hyperparameter tuning was conducted empirically to balance solution quality and runtime. A high initial temperature ($T_0 = 1000.0$) was selected to enhance exploration in the early search stages. To further investigate the robustness of our simulated annealing approach, we conducted a sensitivity analysis on the cooling rate parameter α, which controls the rate of temperature decay. The cooling rate directly influences the exploration-exploitation trade-off during optimization, impacting both convergence quality and runtime. We tested three different values: $\alpha = 0.85$, 0.90, and 0.95 on the Adult dataset with $k = 10$ and $Supp_{max} = 30\%$ (see Table 3). As shown, a higher cooling rate (i.e., slower temperature decay) leads to better classification performance, as it allows the algorithm to explore a larger portion of the solution space and avoid premature convergence to local optima. The improvement, however, diminishes as α approaches 0.95, indicating diminishing returns from slower cooling beyond a certain threshold. This behavior highlights the importance of selecting an appropriate cooling schedule. While $\alpha = 0.95$ yields slightly better accuracy, it also increases the number of iterations required for convergence, thus incurring higher computational cost. On the other hand, lower values such as $\alpha = 0.85$ result in faster convergence but risk settling into suboptimal solutions due to limited exploration. In our experiments, we found that setting $\alpha = 0.90$ achieves a good balance between exploration and convergence speed. When combined with a sufficiently high initial temperature ($T_0 = 1000$) and a fixed number of iterations per temperature level (10), this configuration ensures that the algorithm can escape poor local optima in early stages while progressively refining the solution. The termination was enforced either when the temperature fell below $T_{\min} = 0.001$ or no improvement was observed over 80 evaluations. These choices were informed by observing trade-offs between suppression compliance and utility scores, where aggressive cooling or low T_0 led to premature convergence, while more permissive settings improved the AUC-ROC but risked violating suppression constraints. This setting maintained suppression levels within the specified $Supp_{max}$ constraint and yielded stable AUC-ROC scores across multiple runs. Therefore, $\alpha = 0.90$ is recommended for medium-sized datasets where both utility preservation and runtime are critical considerations. These results highlight a fundamental trade-off between optimization depth and efficiency: while slower cooling (higher α) enhances accuracy, it also incurs greater computational cost. For time-sensitive applications, $\alpha = 0.90$ offers a balanced compromise.

5.3 Experimental Results

We analyzed the effects of anonymity levels and maximum suppression limits by varying the maximum suppression limit ($Supp_{max}$) and the k-anonymity level on classification performance. $Supp_{max}$ was set to 10%, 20%, and 30%, while k was varied between 2 and 20. Figure 2 shows the AUC-ROC scores achieved across different settings. In the MIMIC-III dataset, for $Supp_{max} = 10\%$ at $k = 10$ and $k = 15$, and for $Supp_{max} = 20\%$ at $k = 15$, feasible anonymized solutions could not be generated, resulting in missing points. This occurs because strict suppression constraints limit feasible generalizations under high k requirements. Generally, as k increases, stricter anonymization requirements lead to greater generalization, resulting in a modest decline in AUC-ROC. However, on the Adult dataset, a slight AUC-ROC improvement is observed from $k = 10$ to $k = 15$, possibly due to the smoothing of noisy features during generalization. Additionally, relaxing the maximum suppression limit yields better AUC-ROC scores: fewer attributes need generalization, preserving discriminative patterns.

We next compared the performance of our method against ARX, using $Supp_{max} = 30\%$ for both datasets. Figure 3 illustrates the AUC-ROC scores for different k values. Our method consistently outperforms ARX across all evaluated k values. For example, at $k = 10$ on the Adult dataset, our approach

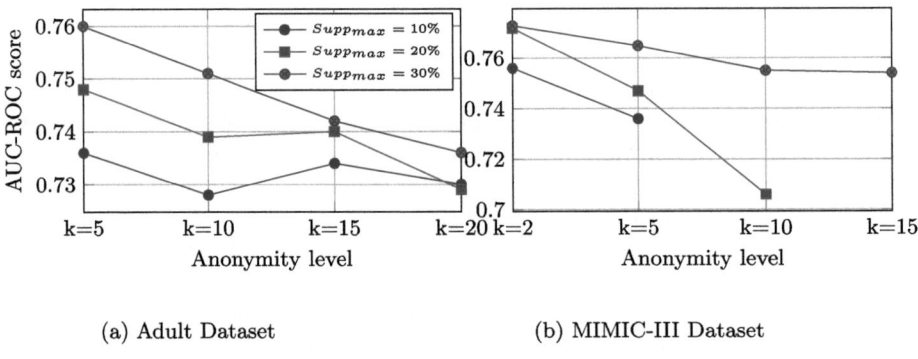

Fig. 2. AUC-ROC under varying suppression limits and k-anonymity levels

achieves an AUC-ROC score of 0.745, compared to 0.721 for ARX—a relative improvement of 3.3%. In the MIMIC-III dataset, at $k = 10$, our method achieves 0.736 compared to 0.702 with ARX. Moreover, in the MIMIC-III dataset at higher privacy settings (e.g., $k = 15$), ARX fails to find a feasible anonymized solution under the suppression constraint, while our method successfully generates valid outputs maintaining acceptable classification performance. This failure mode highlights the limitations of rule-based anonymization in handling stricter constraints, which our method overcomes via flexible search; our method better preserves discriminative patterns even under tight privacy constraints, showing its suitability for clinical applications where both utility and privacy are critical.

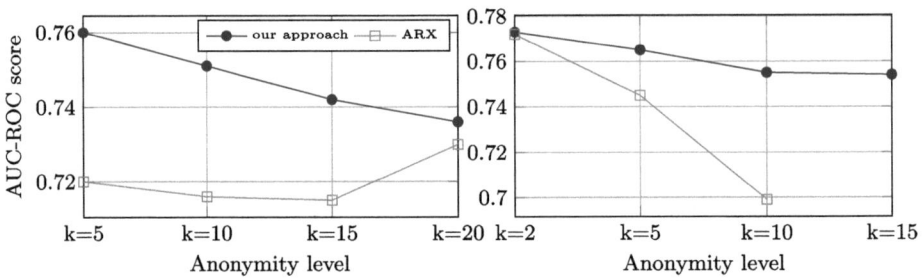

Fig. 3. AUC-ROC scores for Adult (left) and MIMIC-III (right)

6 Conclusions and Future Work

We introduced a novel k-anonymization method leveraging simulated annealing algorithm over generalization hierarchies, combined with suppression, to identify near-optimal anonymization solutions maximizing classification utility. Our approach systematically explores the anonymization search space, probabilistically avoiding local optima to find high-quality solutions. Extensive experiments on real-world datasets show that our method consistently outperforms the state-of-the-art ARX framework in classification accuracy post-anonymization. For future work, we plan to optimize the construction of generalization hierarchies, investigating hybrid metaheuristics (e.g., combining simulated annealing with genetic algorithms), and broadening applicability to alternative privacy models such as l-diversity and t-closeness. We also aim to explore incremental evaluation strategies to further improve computational efficiency.

Acknowledgement. This work was funded by the German Federal Ministry of Education and Research and NextGenerationEU by the European Union under grant number 16KISA001K (PrivacyUmbrella).

Data Availability Statement. The source code used in this study is available at: https://github.com/desstaw/PrivacyPreservingTechniques.

Disclosure of Interests. The authors have no competing interests.

References

1. Ayala-Rivera, V., McDonagh, P., Cerqueus, T., Murphy, L., et al.: A systematic comparison and evaluation of k-anonymization algorithms for practitioners. Trans. Data Priv. **7**(3), 337–370 (2014)
2. Bayardo, R.J., Agrawal, R.: Data privacy through optimal k-anonymization. In: 21st International Conference on Data Engineering (ICDE'05), pp. 217–228. IEEE (2005)
3. Becker, B., Kohavi, R.: Adult. UCI Machine Learning Repository (1996). https://doi.org/10.24432/C5XW20
4. Caruccio, L., Desiato, D., Polese, G., Tortora, G., Zannone, N.: A decision-support framework for data anonymization with application to machine learning processes. Inf. Sci. **613**, 1–32 (2022)
5. Dorogush, A.V., Ershov, V., Gulin, A.: Catboost: gradient boosting with categorical features support. arXiv preprint arXiv:1810.11363 (2018)
6. Dwork, C., McSherry, F., Nissim, K., Smith, A.: Calibrating noise to sensitivity in private data analysis. In: Halevi, S., Rabin, T. (eds.) TCC 2006. LNCS, vol. 3876, pp. 265–284. Springer, Heidelberg (2006). https://doi.org/10.1007/11681878_14
7. Emam, K., et al.: A globally optimal k-anonymity method for the de-identification of health data. J. Am. Med. Inform. Assoc. JAMIA **16**, 670–682 (2009). https://doi.org/10.1197/jamia.M3144
8. Ge, Y.F., Wang, H., Cao, J., Zhang, Y., Jiang, X.: Privacy-preserving data publishing: an information-driven distributed genetic algorithm. World Wide Web **27** (2024). https://doi.org/10.1007/s11280-024-01241-y
9. Iyengar, V.S.: Transforming data to satisfy privacy constraints. In: Proceedings of the Eighth ACM SIGKDD International Conference on Knowledge Discovery and Data Mining, pp. 279–288 (2002)
10. Johnson, A., Pollard, T., Mark, R.: MIMIC-III Clinical Database (version 1.4) (2016). https://doi.org/10.13026/C2XW26
11. Kohlmayer, F., Prasser, F., Eckert, C., Kemper, A., Kuhn, K.A.: Flash: efficient, stable and optimal k-anonymity. In: 2012 International Conference on Privacy, Security, Risk and Trust and 2012 International Confernece on Social Computing, pp. 708–717. IEEE (2012)

12. LeFevre, K., DeWitt, D.J., Ramakrishnan, R.: Incognito: efficient full-domain k-anonymity. In: Proceedings of the 2005 ACM SIGMOD International Conference on Management of Data, pp. 49–60 (2005)
13. Li, N., Li, T., Venkatasubramanian, S.: t-closeness: Privacy beyond k-anonymity and l-diversity. In: 2007 IEEE 23rd International Conference on Data Engineering, pp. 106–115. IEEE (2006)
14. Machanavajjhala, A., Kifer, D., Gehrke, J., Venkitasubramaniam, M.: l-diversity: Privacy beyond k-anonymity. ACM Trans. Knowl. Discov. Data (TKDD) **1**(1), 3-es (2007)
15. Meyerson, A., Williams, R.: On the Complexity of Optimal k-Anonymity, vol. 23, pp. 223–228 (2004). https://doi.org/10.1145/1055558.1055591
16. Ni, C., Cang, L.S., Gope, P., Min, G.: Data anonymization evaluation for big data and IoT environment. Inf. Sci. **605**, 381–392 (2022)
17. Samarati, P.: Protecting respondents identities in microdata release. IEEE Trans. Knowl. Data Eng. **13**(6), 1010–1027 (2001)
18. Slijepčević, D., Henzl, M., Klausner, L.D., Dam, T., Kieseberg, P., Zeppelzauer, M.: k-anonymity in practice: how generalisation and suppression affect machine learning classifiers. Comput. Secur. **111**, 102488 (2021)
19. Su, B., Huang, J., Miao, K., Wang, Z., Zhang, X., Chen, Y.: K-anonymity privacy protection algorithm for multi-dimensional data against skewness and similarity attacks. Sensors **23**(3), 1554 (2023)
20. Sweeney, L.: Achieving k-anonymity privacy protection using generalization and suppression. Internat. J. Uncertain. Fuzziness Knowl.-Based Syst. **10**(05), 571–588 (2002)
21. Sweeney, L.: k-Anonymity: a model for protecting privacy. Internat. J. Uncertain. Fuzziness Knowl.-Based Syst. **10**(05), 557–570 (2002)
22. Wang, R., Zhu, Y., Chang, C.C., Peng, Q.: Privacy-preserving high-dimensional data publishing for classification. Comput. Secur. **93**, 101785 (2020)
23. Winkler, W.: Using simulated annealing for k-anonymity (2002)
24. Zigomitros, A., Casino, F., Solanas, A., Patsakis, C.: A survey on privacy properties for data publishing of relational data. IEEE Access **8**, 51071–51099 (2020)

Predicting Gelation in Copolymers Using Deep Learning Through a Comparative Study of ANN, CNN, and LSTM Models with SHAP Explainability

Selahattin Barış Çelebi[1(✉)], Ammar Aslan[2], and Mutlu Canpolat[3]

[1] Visiting Research Fellow: Department of Computer Science, University of Warwick, Coventry, UK
sbariscelebi@gmail.com
[2] Department of Computer Technologies, Vocational School of Technical Sciences, Batman University, 72100 Batman, Turkey
[3] Department of Chemistry and Chemical Processing Technologies, Vocational School of Technical Sciences, Batman University, 72100 Batman, Turkey

Abstract. This study presents a Deep Learning (DL)-based approach to predict gelation behavior in copolymer systems using compositional and physicochemical descriptors. Three architectures—Artificial Neural Network (ANN), Convolutional Neural Network (CNN), and Long Short-Term Memory (LSTM)—were tested and evaluated under conditions of pronounced class imbalance. The ANN model achieved the best performance, with an Accuracy (ACC) of 94% and an F1-score of 0.57, demonstrating strong discriminative capability in the binary classification of gelation propensity. To enhance robustness, threshold optimization was employed, and SHapley Additive exPlanations (SHAP) was used to identify key predictors, including specific monomer concentrations and degree of polymerization. The findings demonstrate that data-driven methods can effectively capture complex gelation patterns and provide interpretable, mechanistically relevant insights. This study underscores the potential of Artificial Intelligence (AI) to accelerate polymer design while reducing reliance on empirical experimentation.

Keywords: Convolutional Neural Network · Long Short-Term Memory · Artificial Neural Network · Polymer · Renewable Energy Materials

1 Introduction

Elucidating the physical and chemical behaviors of polymers remains a cornerstone of contemporary materials science. Among these behaviors, gelation represents a particularly intricate phenomenon, wherein polymers undergo a sol–gel transition by forming three-dimensional network structures. This process is critical for engineering advanced functional materials, with applications ranging from biomedical devices to high-performance composites and renewable energy materials [1, 2]. Despite its significance, accurately predicting gelation in polymer systems solely through experimental

© The Author(s), under exclusive license to Springer Nature Switzerland AG 2026
G. Bergami et al. (Eds.): IDEAS 2025, LNCS 15928, pp. 85–95, 2026.
https://doi.org/10.1007/978-3-032-06744-9_7

approaches remains a formidable challenge, owing to the multitude of interrelated factors and the inherent nonlinearity of their interactions [3]. Recent studies have explored the synthesis and cross-linking behavior of functional copolymers in dye and heavy metal removal applications, which inherently rely on gelation characteristics and network formation dynamics, thereby underscoring the broader relevance of gelation analysis in polymer design [4, 5].

Machine Learning (ML) algorithms-based studies have primarily focused on environmental factors or molecular structure, with limited attention paid to the role of compositional descriptors in gelation outcomes. To the best of our knowledge, no prior work has directly formulated gelation prediction as a binary classification task using only polymer composition and physicochemical features as inputs. To address this gap, the present study applies ANN, CNN, and LSTM models to predict gelation behavior directly from compositional and physicochemical descriptors. Furthermore, by leveraging Explainable Artificial Intelligence (XAI) techniques—SHapley Additive exPlanations (SHAP)—this study systematically identifies and quantifies the contributions of individual monomers and physicochemical variables to gelation propensity, thereby providing interpretable insights into the underlying chemical mechanisms.

This study provides a complementary modeling approach to experimental workflows in polymer design, with the specific objectives of developing ML models for gelation prediction, comparing model performance, and establishing an interpretable, data-driven framework to aid material design.

2 Related Study Overview

In recent years, ML techniques have achieved remarkable success across a diverse range of scientific and engineering fields [6], including image analysis [7], natural language processing [8], and information security [9]. However, despite their transformative impact, the application of ML within polymer science remains comparatively limited, hindered by challenges such as scarce, heterogeneous datasets and complex polymer representations, despite some recent advancements [10, 11]. For instance, Huang et al. (2024) applied linear regression (LR), support vector regression (SVR), Decision Tree Regressor (DTR), Random Forest Regression (RFR), and Deep Neural Network (DNN) to predict the gelation in conductive gelatin methacryloyl hydrogels. Using SVR, their approach obtained a Coefficient of Determination (R2) of 0.79 and a Mean Absolute Percentage Error (MAPE) of 3.13% for the prediction of polymer melt viscosity [12]. Similarly, Jain et al. (2025) developed a Physics-Enforced Neural Network (PENN) to estimate the melt viscosity of thermoplastic systems, achieving a R2 of 0.79. This model integrated domain-specific knowledge directly into its architecture [13]. While these and other studies—such as those by Perera et al. (2025) [14] and Bemani et al. (2023) [15]— demonstrate the potential of ML in polymer informatics, they predominantly address continuous-valued properties such as viscosity, gelation, or texture. Most of these efforts rely on regression-based modeling frameworks. However, the direct classification of gelation behavior—specifically, determining whether a given polymer composition will result in gelation—remains underexplored in the existing literature [16, 17]. To bridge this gap, the current study formulates gelation prediction as a binary classification problem. We propose and evaluate three DL architectures: multilayer perceptron (MLP-ANN), CNN, and LSTM networks-to classify gelation tendency based on compositional descriptors. Furthermore, the proposed approach incorporates threshold calibration, post hoc explainability using SHAP, and ablation analysis to enable a comprehensive evaluation of model performance and interpretability.

3 Material and Methods

The Horseradish Peroxidase (HRP) dataset was obtained from the Princeton University Data Commons (PUDC), an open-access institutional repository for scientific purposes. It includes the following features: Polymer Number (PN), Degree of Polymerization (DP), 2-(Diethylamino) Ethyl Methacrylate (DEAEMA), 2-Hydroxypropyl Methacrylate (HPMA), 2-Sulfopropyl Methacrylate (SPMA), Butyl Methacrylate (BMA), 3-(Dimethylamino) Propyl Methacrylate (DMAPMA), Methyl Methacrylate (MMA), Poly (Ethylene Glycol) Methyl Ether Methacrylate (PEGMA), Trimethylammonium Chloride Ethyl Methacrylate (TMAEMA), Retained Enzyme Activity (REA), Generation (Experimental round/Active learning cycle), and Gelation (Binary outcome: 0/1). HRP serves to evaluate the stability of polymer–protein hybrids under heat treatment. During model development, the features PN (identifier), REA (measured after synthesis and enzymatic assay), and Generation (indicating the synthesis round within the active learning cycle) were excluded from model inputs due to their lack of predictive relevance for the ML algorithm. The Gelation label (0: 588, 1: 36; ~5.8% positive cases) was designated as the target. To address the imbalance challenge, a comprehensive DL pipeline was established, encompassing data preprocessing, model training, threshold calibration, SHAP-based explainability, and ablation-based performance validation [18], as illustrated in Fig. 1.

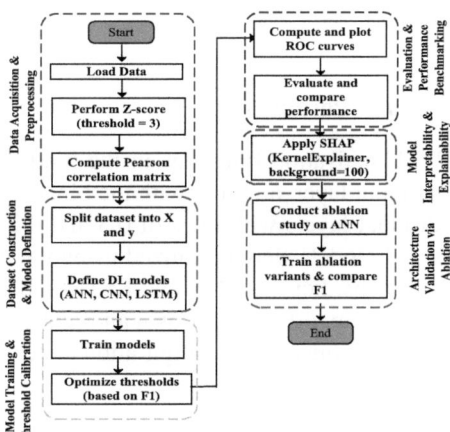

Fig. 1. Schematic overview of the DL pipeline for polymer gelation prediction.

3.1 Data Preprocessing

To enhance model robustness and reduce the influence of statistical outliers, z-score-based outlier detection was employed [19]. Figure 2 illustrates the distributions of numerical features in the HRP dataset prior to the preprocessing step. Several features—including DEAEMA and PEGMA—displayed pronounced skewness and heavy-tailed characteristics at this stage. For example, PEGMA exhibited a right-skewed distribution, indicating a substantial presence of high-value outliers. Figure 3 presents the same feature distributions following the application of z-score-based outlier removal. This preprocessing step led to a marked reduction in skewness and tail heaviness for the affected features. Most distributions became more symmetric, and extreme outliers were effectively eliminated, resulting in improved data normality and enhanced suitability for subsequent ML analyses.

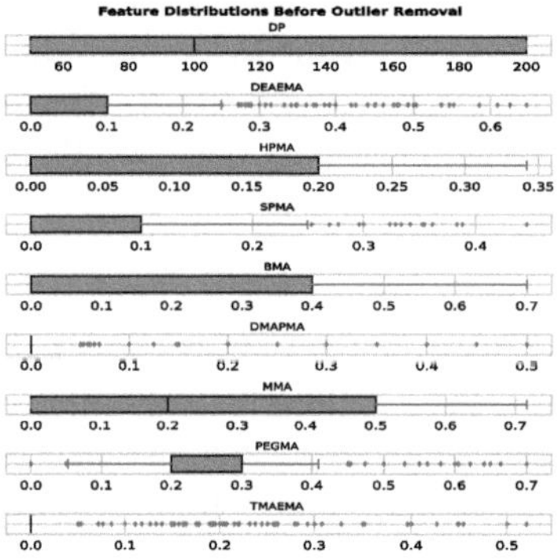

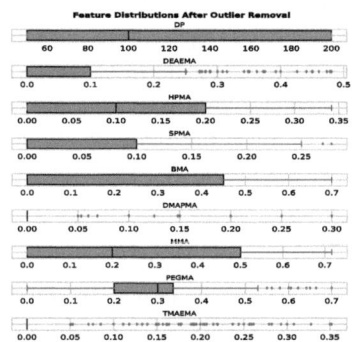

Fig. 2. HRP dataset before preprocessing.

Fig. 3. HRP dataset after preprocessing.

3.2 Model Architectures and Parameters

All models employed ReLU activation functions in the hidden layers and a sigmoid activation function in the output layer, aligning with the binary classification objective. To mitigate overfitting, dropout layers with rates of 0.3 and 0.2 were applied between dense or recurrent layers. All models were optimized using the Adam optimizer with a batch size of 16 over 150 epochs. For the CNN and LSTM models, input tensors were reshaped to a shape of (9, 1) to support convolutional and sequential processing. Table 1 provides an overview of the architectural configurations and corresponding training hyperparameters.

Table 1. Model architectures and training hyperparameters used for binary gelation classification.

Model	Layers	Dropout	Activation	Optimizer	Input Shape
ANN	[64, 32, 16, 1]	0.3/0.2	ReLU + Sigmoid	Adam	(9,)
CNN	[64C, 32C, FC32, 1]	0.3/0.2	ReLU + Sigmoid	Adam	(9,1)
LSTM	[LSTM64, LSTM32, FC16, 1]	0.3/0.2	Tanh + Sigmoid (LSTM), ReLU, Sigmoid	Adam	(9,1)

In Table 1, the notation "C" refers to one-dimensional convolutional layers, while "FC" denotes fully connected (dense) layers. CNN and LSTM models were provided with reshaped inputs of shape (9, 1) to facilitate convolutional and sequential computations.

3.3 Training Strategy

The dataset was randomly partitioned into training (64%), validation (16%), and test (20%) sets using a fixed random seed (random_state = 42) to ensure reproducibility. To mitigate the effects of class imbalance, class weights inversely proportional to class frequencies were incorporated during model training, enhancing the model's sensitivity to minority-class instances. Model performance was assessed using five standard classification metrics: Accuracy (ACC), precision (PRE), Recall, F1-score, and the Area Under the Receiver Operating Characteristic curve (ROC-AUC) [20, 21].

3.4 Explainability with SHAP

To enhance model interpretability, SHAP was employed to quantify the marginal contribution of each input feature to the model's output [22]. The KernelExplainer was used due to its model-agnostic design. Background samples were randomly drawn from the training set (n = 100), and SHAP values were computed for both local (n = 1) and global (n = 50) instances. Summary plots were used to identify globally influential features, while force plots facilitated localized interpretation of specific gelation predictions.

4 Results and Discussion

This section presents and analyzes the performance of the DL models in the context of gelation prediction in copolymer systems. This evaluation encompasses classification metrics, AUC analysis, threshold optimization, SHAP-based feature importance, and an ablation study to assess the impact of architectural components.

4.1 Model Performance Evaluation

The feedforward ANN model achieved the highest F1-score (0.57) and an AUC = 0.83 on the test set, indicating a favorable precision–recall trade-off despite class imbalance. Table 2 summarizes the classification metrics of the ANN, CNN, and LSTM models on both the training and test datasets. The ANN achieved the highest test accuracy (0.94), compared to CNN (0.87) and LSTM (0.92), supporting its effectiveness. Although CNN and LSTM achieved recall values of 0.66 and 0.50 on the test set, their relatively low precision scores (0.25 and 0.37, respectively) resulted in reduced F1-scores (0.36 and 0.42, respectively). The ANN model demonstrated stable performance across the training (F1 = 0.59) and test sets (F1 = 0.57), suggesting valuable generalization capability. Based on these performance metrics, the ANN captures discriminative patterns more effectively than sequential models (CNN and LSTM) in this binary gelation classification task. The performance of the ANN can be attributed to the nature of the input data, which comprises independent physicochemical and compositional descriptors without inherent sequential or spatial dependencies. Prior studies have demonstrated that for real-world tabular datasets, DNNs, including CNNs and LSTMs, do not necessarily outperform simpler feedforward architectures or classical ML models [23, 24].

Table 2. Classification performance metrics for ANN, CNN, and LSTM models evaluated on training and test dataset.

Model	Dataset	Accuracy	Precision	Recall	F1-Score
ANN	Train	0.94	0.51	0.70	0.59
CNN	Train	0.91	0.36	0.83	0.50
LSTM	Train	0.93	0.43	0.54	0.48
ANN	Test	0.94	0.50	0.66	0.57
CNN	Test	0.87	0.25	0.66	0.36
LSTM	Test	0.92	0.37	0.50	0.42

4.2 Comparative ROC and Threshold Optimization Analysis

The discriminative performance of each model was assessed using ROC analysis. The LSTM model achieved the lowest AUC (0.74), the CNN yielded a higher AUC (0.78), and the ANN achieved the highest AUC score (0.83) (see Fig. 4a–c). This metric is particularly important in imbalanced classification settings. The ANN architecture delivered consistent results across all evaluation metrics, supporting its selection as the most robust model in this study.

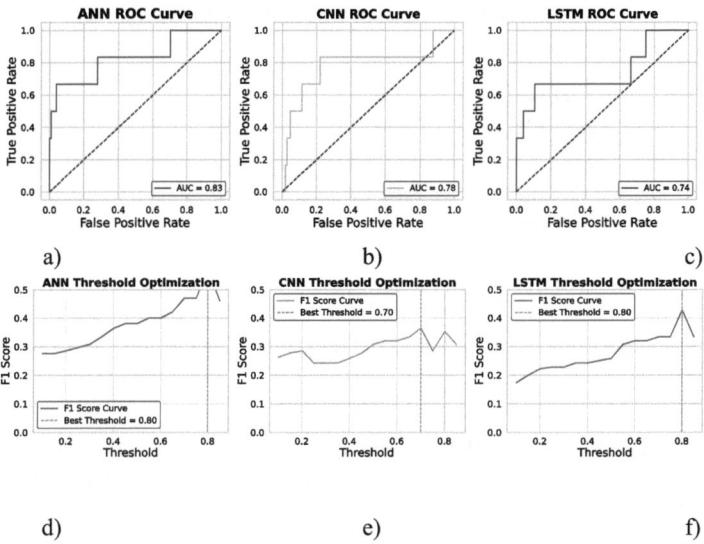

Fig. 4. Comparative ROC curves for the ANN (a), CNN (b), and LSTM (c) models with F1-score threshold optimization curves for the ANN (d), CNN (e), and LSTM (f) models.

In Fig. 4d–f, the optimal thresholds were 0.80 for the ANN, 0.70 for the CNN, and 0.80 for the LSTM. Threshold calibration involved maximizing the F1-score across probability thresholds ranging from 0.1 to 0.9 with a step size of 0.05. As shown in Fig. 5d, the F1-score curve for the ANN plateaued at a high value, indicating robust calibration and effective class separation.

4.3 Feature Importance Analysis with SHAP

Figure 5 displays the global feature importance for gelation prediction, quantified by SHAP values. Among all features, PEGMA exhibits the greatest influence on the model's output, with higher PEGMA values strongly associated with an increased likelihood of gelation. HPMA and DEAEMA exhibit moderate contributions, while BMA and DP have less significant but measurable influence. The exclusion of the generation feature aligns with the preprocessing strategy, which omitted non-predictive or post hoc data from model inputs. These results indicate that specific monomer compositions, particularly PEGMA, primarily determine gelation, supporting the structure–property relationships established through polymer cross-linking chemistry [25].

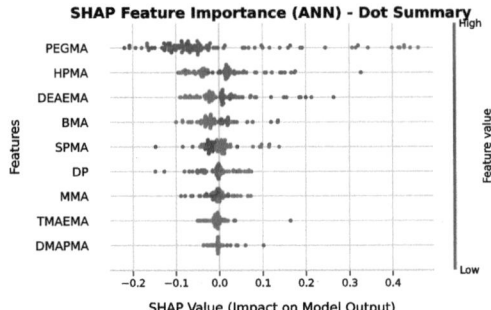

Fig. 5. SHAP summary plot for the ANN model indicating global feature importance in gelation prediction.

Monomers like PEGMA, HPMA, and DEAEMA can directly affect gelation by changing the hydrophilic–hydrophobic balance and crosslinking potential between polymer chains. PEGMA's long flexible chains and the functional groups in HPMA and DEAEMA may help form a network and increase water retention [25].

4.4 ANN Architecture Ablation Study

As summarized in Table 3, removing dropout led to a notable decrease in performance, reducing the F1-score from 0.57 to 0.42. In addition, eliminating the middle hidden layer or further simplifying the model to a single hidden layer both resulted in an F1-score of 0.34. These findings indicate that dropout, under the present data and model conditions, prevents overfitting and that a single hidden layer is insufficient to improve predictive performance.

Table 3. Performance impact of architectural modifications to the ANN model.

ANN Variant	Dropout	Hidden Layers	F1-Score
Base ANN	Yes	[64, 32, 16]	0.57
No Dropout	No	[64, 32, 16]	0.42
Mid Layer Removed	Yes	[64, 16]	0.34
Fewer Layers (Simplified)	No	[64]	0.34

4.5 Discussion

Recent ML applications in polymer science have mainly focused on regression tasks, such as predicting gel fraction and melt viscosity. However, classification of gelation behavior, particularly with class imbalance, remains underexplored. This study addresses this gap and invites future comparison as the literature evolves. Although transformer-based models show promise, recent studies suggest their advantages over traditional neural networks are not consistent for small- to medium-sized tabular data [24]. Future work will investigate advanced architectures as data size grows. Another limitation is the manual design of model architecture, which may affect generalizability. While ablation and comparative analyses systematically assessed architectural impact, model performance could be improved by advanced optimization methods like Optuna with TPE and random search [26].

5 Conclusion

This study introduced DL models for predicting and analyzing the gelation behavior in copolymer systems using compositional and physicochemical descriptors. Among the evaluated architectures, the ANN achieved the highest performance, with an ACC of 0.94 and a 36% improvement in F1-score over the CNN model and a 58% improvement over the LSTM model, while also exhibiting a more balanced precision–recall trade-off despite severe class imbalance. Threshold optimization enhanced classification robustness, particularly in detecting minority-class gelation cases, whereas SHAP-based explainability identified key predictive features, including specific monomer types and the degree of polymerization. These findings provide not only a reliable in silico tool for gelation prediction but also mechanistic insights into the factors governing cross-linking behavior. Overall, the results underscore the potential of data-driven approaches in streamlining polymer design and reducing reliance on time-consuming experimental procedures. Future work could focus on incorporating additional data or advanced architectures to improve model generalizability and alignment with the complex dynamics of real-world polymerization systems.

Code Availability. he complete implementation of the deep learning pipeline is publicly available on GitHub: https://github.com/sbariscelebi/gelation-prediction-ml.

Disclosure of Interests. The authors have no competing interests to declare that are relevant to the content of this article.
Data Availability. The dataset used in this study, comprising 624 copolymer samples with compositional and physicochemical features, is publicly available at [18].

References

1. Bommarius, A.S., Paye, M.F.: Stabilizing biocatalysts. Chem. Soc. Rev. **42**, 6534–6565 (2013). https://doi.org/10.1039/C3CS60137D
2. Panganiban, B., et al.: Random heteropolymers preserve protein function in foreign environments. Science **359**, 1239–1243 (2018). https://doi.org/10.1126/science.aao0335
3. Mechanics of Soft Gels: linear and nonlinear response. In: Handbook of Materials Modeling, pp. 1719–1746. Springer International Publishing, Cham (2020). https://doi.org/10.1007/978-3-319-44680-6_129
4. Canpolat, M., Topal, G.: Synthesis, characterization of cross-linked poly(ethylene glycol) dimethacrylate-methyl methacrylate-N-(1-phenylethyl) acrylamide) copolymer and removal of copper(II), cobalt(II) ions from aqueous solutions via this copolymer. Environ. Prog. Sustain. Energy **42**, e14197 (2023). https://doi.org/10.1002/ep.14197

5. Sayar, M., Canpolat, M.: Novel Poly(N-(1-cyclohexylethyl)Acrylamide-co-Styrene-co-Ethyleneglycol Dimethacrylate) synthesis for efficient methylene blue adsorption from aqueous solutions synthesis. ChemistrySelect **9**, e202400947 (2024). https://doi.org/10.1002/slct.202400947
6. Çelebġ, S.B., Karaman, Ö.A., Öztürk, G.: Autonomous driving technologies and cyber security in vehicles (2022)
7. Aslan, A., Çelebi, S.B.: Real time deep learning based age and gender detection for advertising and marketing. Uluslar. Bilişim Kongresi IIC 2022 Bildir. Kitabı. 10–16 (2022)
8. Sunar, A.S., Khalid, M.S.: Natural language processing of student's feedback to instructors: a systematic review. IEEE Trans. Learn. Technol. **17**, 741–753 (2024). https://doi.org/10.1109/TLT.2023.3330531
9. Kılıç, İ., Yaman, O., Erdoğan, E., Aslan, M.İ.: A hybrid method based on a genetic algorithm that uses network packets to classify spyware. J. Phys. Chem. Funct. Mater. **7**, 148–157 (2024). https://doi.org/10.54565/jphcfum.1579687
10. Malashin, I., Tynchenko, V., Gantimurov, A., Nelyub, V., Borodulin, A.: Boosting-based machine learning applications in polymer science: a review. Polymers **17**, 499 (2025). https://doi.org/10.3390/polym17040499
11. Struble, D.C., Lamb, B.G., Ma, B.: A prospective on machine learning challenges, progress, and potential in polymer science. MRS Commun. **14**, 752–770 (2024). https://doi.org/10.1557/s43579-024-00587-8
12. Huang, X., Wong, Y.X., Goh, G.L., Gao, X., Lee, J.M., Yeong, W.Y.: Machine learning-driven prediction of gel fraction in conductive gelatin methacryloyl hydrogels. Int. J. AI Mater. Des. **1**, 61 (2024). https://doi.org/10.36922/ijamd.3807
13. Jain, A., Gurnani, R., Rajan, A., Qi, H.J., Ramprasad, R.: A physics-enforced neural network to predict polymer melt viscosity. NPJ Comput. Mater. **11** (2025). https://doi.org/10.1038/s41524-025-01532-6
14. Perera, Y.S., Li, J., Abeykoon, C.: Machine learning enhanced grey box soft sensor for melt viscosity prediction in polymer extrusion processes. Sci. Rep. **15** (2025). https://doi.org/10.1038/s41598-025-85619-6
15. Bemani, A., Madani, M., Kazemi, A.: Machine learning-based estimation of nano-lubricants viscosity in different operating conditions. Fuel **352**, 129102 (2023). https://doi.org/10.1016/j.fuel.2023.129102
16. Mencucci, R., Cennamo, M., Favuzza, E., Rechichi, M., Rizzo, S.: Triphasic polymeric corneal coating gel versus a balanced salt solution irrigation during cataract surgery: a postoperative anterior segment optical coherence tomography analysis and confocal microscopy evaluation. J Cataract Refract Surg **45**, 1148–1155 (2019). https://doi.org/10.1016/j.jcrs.2019.03.002
17. Rangchian, A., Hubschman, J.-P., Kavehpour, H.P.: Time dependent degradation of vitreous gel under enzymatic reaction: polymeric network role in fluid properties. J. Biomech. **109**, 109921 (2020). https://doi.org/10.1016/j.jbiomech.2020.109921
18. Webb, M., Patel, R., Gormley, A., Tamasi, M., Borca, C., Kosuri, S.: Data on Enzyme Activity Retention in glucose oxidase, lipase, and horseradish peroxidase, https://datacommons.princeton.edu/discovery/doi/https/doi.org/10.34770/h938-nn26 (2022).
19. Aggarwal, V., Gupta, V., Singh, P., Sharma, K., Sharma, N.: Detection of spatial outlier by using improved z-score test. In: 2019 3rd International Conference on Trends in Electronics and Informatics (ICOEI), pp. 788–790. IEEE, Tirunelveli, India (2019). https://doi.org/10.1109/icoei.2019.8862582
20. Krstinić, D., Braović, M., Šerić, L., Božić-Štulić, D.: Multi-label classifier performance evaluation with confusion matrix. Comput. Sci. Inf. Technol. **1**, 1–14 (2020)
21. Fawcett, T.: An introduction to ROC analysis. Pattern Recognit. Lett. **27**, 861–874 (2006). https://doi.org/10.1016/j.patrec.2005.10.010

22. Pelegrina, G.D., Duarte, L.T., Grabisch, M.: A k-additive Choquet integral-based approach to approximate the SHAP values for local interpretability in machine learning. https://arxiv.org/abs/2211.02166 (2022). https://doi.org/10.48550/ARXIV.2211.02166
23. Kadra, A., Lindauer, M., Hutter, F., Grabocka, J.: Regularization is all you need: Simple neural nets can excel on tabular data. ArXiv Prepr. ArXiv210611189, vol. 536 (2021)
24. Shwartz-Ziv, R., Armon, A.: Tabular data: deep learning is not all you need. Inf. Fusion. **81**, 84–90 (2022). https://doi.org/10.1016/j.inffus.2021.11.011
25. Hwang, J.W., Noh, S.M., Kim, B., Jung, H.W.: Gelation and crosslinking characteristics of photopolymerized poly(ethylene glycol) hydrogels. J. Appl. Polym. Sci. **132**, app. 41939 (2015). https://doi.org/10.1002/app.41939
26. Akiba, T., Sano, S., Yanase, T., Ohta, T., Koyama, M.: Optuna: A next-generation hyperparameter optimization framework. In: Proceedings of the 25th ACM SIGKDD International Conference on Knowledge Discovery & Data Mining, pp. 2623–2631. ACM, Anchorage AK USA (2019). https://doi.org/10.1145/3292500.3330701

A Total Variation Regularized Framework for Epilepsy-Related MRI Image Segmentation

Mehdi Rabiee[ID], Sergio Greco[ID], Reza Shahbazian[ID], and Irina Trubitsyna(✉)[ID]

Department of Computer Engineering, Modeling, Electronics and Systems (DIMES), University of Calabria, 87036 Rende, Italy
i.trubitsyna@dimes.unical.it

Abstract. Focal Cortical Dysplasia (FCD) is a primary cause of drug-resistant epilepsy and is difficult to detect in brain magnetic resonance imaging (MRI) due to the subtle and small-scale nature of its lesions. Accurate segmentation of FCD regions in 3D multimodal brain MRI images is essential for effective surgical planning and treatment. However, this task remains highly challenging due to the limited availability of annotated FCD datasets, the extremely small size and weak contrast of FCD lesions, the complexity of handling 3D multimodal inputs, and the need for output smoothness and anatomical consistency, which is often not addressed by standard voxel-wise loss functions. This paper presents a new framework for segmenting FCD regions in 3D brain MRI images. We adopt state-of-the-art transformer-enhanced encoder-decoder architecture and introduce a novel loss function combining Dice loss with an anisotropic Total Variation (TV) term. This integration encourages spatial smoothness and reduces false positive clusters without relying on post-processing. The framework is evaluated on a public FCD dataset with 85 epilepsy patients and demonstrates superior segmentation accuracy and consistency compared to standard loss formulations. The model with the proposed TV loss shows an 11.9% improvement on the Dice coefficient and 13.3% higher precision over the baseline model. Moreover, the number of false positive clusters is reduced by 61.6%.

Keywords: Image Segmentation · 3D MRI · Deep Learning · Focal Cortical Dysplasia · Medical Data

1 Introduction

Epilepsy is a neurological disorder characterized by a persistent predisposition to generate unprovoked seizures, affecting millions worldwide and necessitating accurate diagnosis and management due to its potential long-term impact on quality of life and brain function [2]. The 75th World Health Assembly and World Health Organization (WHO) selected epilepsy as one of the top priorities in the prevention and control of noncommunicable diseases [1].

Epilepsy is often linked to lesions or abnormalities on the brain's cortex, which trigger and spread seizures. The most common cause is focal cortical dysplasia (FCD), which encompasses a spectrum of developmental malformations in the cerebral cortex, marked by localized disruptions in cortical architecture and cellular composition [13]. It is considered the leading cause of drug-resistant epilepsy in children and remains a significant factor in the use of anti-epileptic medications among adults [7]. Identifying FCD regions in brain magnetic resonance imaging (MRI) images is vital for successful surgery and a better chance of curing epilepsy.

Artificial intelligence (AI), and in particular deep learning (DL), can help find potential epilepsy regions. The goal is to perform a medical image segmentation task where the inputs are 3D volumes consisting of voxels (similar to pixels in 2D images), usually reconstructed from a sequence of 2D MRI images recorded by medical imaging devices with known position and orientation of the recording device for each frame. 3D images usually have multiple modalities, which are captured with different parameters of the MRI scanner and can get different aspects of brain structure; for example, T1-weighted, T2-weighted, FLAIR, and PET [14]. Although the problem is similar to other medical imaging diagnosis tasks, like detecting brain tumors from MRI images or analyzing 3D medical images from CT scans, detecting FCD regions is more challenging because of their very small sizes that are hard to see even by experts. Another important issue is the availability of robust and annotated training data. Unlike many other tasks, there are only a few small-sized datasets available for FCD detection. Therefore, it is essential to utilize a robust architecture with optimal hyperparameters and training strategies to achieve the best possible results.

Medical image segmentation is a well-studied subject. The state of the art for medical image segmentation is based on *U-Net* architecture [10]. This architecture is shaped like a letter *U* that consists of a symmetric encoder-decoder architecture with a contracting path for feature extraction and an expansive path for precise localization. A skip connection between the decoder and the corresponding encoder block at each level enables the model to use fine detail information from encoders in the reconstruction path. This design allows *U-Net* to efficiently learn spatial hierarchies and retain high-resolution contextual information, making it especially effective for pixel-wise segmentation tasks in biomedical imaging. With the emergence of transformer architectures and their success in language processing and computer vision tasks, some studies combined the idea of transformers with the well-established *U-Net* architecture. In particular, vision transformers or attention blocks are used to capture long-range dependencies and global context, complementing convolutional features. Some well-performing architectures are *UNETR* [4], *Swin UNETR* [3], *UNETR++* [12], and *MS-DSA-Net* that outperforms the other existing methods in FCD detection task [15].

Contributions. In this paper we focus on a real-world clinical challenge: segmenting FCD regions in 3D brain MRI images, and adopt the state-of-the-art method *MS-DSA-Net* [15] as the base. Due to the limited size and complexity of available FCD datasets, we carefully design a training pipeline based on patch-wise sampling and voxel-wise classification, enabling the model to learn effectively from limited and high-dimensional data. We propose a new loss by adding a Total Variation (TV) regularization term to the loss function, which encourages the model to produce smoother and more anatomically consistent segmentation masks by penalizing abrupt changes in neighboring voxel predictions. We validate our proposed approach on a publicly available dataset of annotated FCD scans. In particular, we consider different combinations of Dice Loss, Cross Entropy Loss and Total Variation component in the presence and absence of post-processing that cleans up noisy or fragmented segmentation outputs. The results show that adding the TV regularization to a standard Dice loss not only improves segmentation accuracy but also leads to cleaner, more coherent prediction maps. This can improve the further advanced AI-based automated detections.

Organization. The rest of this paper is organized as follows: Sect. 2 reviews related works on deep learning models for medical image segmentation, including transformer-based architectures. Section 3 introduces our proposed model, including the incorporation of Total Variation (TV) loss into the MS-DSA-Net architecture. Section 4 describes the experimental setup, dataset, training details, and evaluation metrics. This section also presents quantitative and qualitative results, followed by an in-depth discussion. Finally, Sect. 5 concludes the paper and outlines future research directions.

2 Related Works

In this section we briefly describe the main architectures of 3D medical image segmentation. U-Net [10] was introduced in 2015 and since then has been utilized as the base architecture for state-of-the-art methods. The U-Net consists of a symmetric encoder-decoder architecture with a contracting path for feature extraction and an expansive path for precise localization. The contracting path applies repeated 3×3 convolutions (without padding), each followed by ReLU and 2×2 max pooling with stride 2, doubling the number of feature channels at each step. The expansive path upsamples the feature maps using 2×2 deconvolutions that halve the feature channels, concatenates them with the corresponding cropped feature maps from the encoder, and applies two 3×3 convolutions followed by ReLU. A final channel-wise convolution maps the output to the desired number of classes. SegResNet [9] uses an encoder-decoder convolutional neural network (CNN) architecture with an asymmetrically larger encoder for feature extraction and a smaller decoder for mask reconstruction. To enhance training on limited data, a variational autoencoder (VAE) branch is added at the encoder's endpoint to reconstruct the input image, providing additional regularization and guidance. The encoder is based on ResNet blocks using $3 \times 3 \times 3$ convolutions with *Group Normalization* and ReLU, combined with strided convolutions for downsampling and skip connections for feature preservation. The encoder reduces spatial dimensions while increasing feature depth. The decoder mirrors this structure but uses fewer blocks per level, upsampling features via non-trainable 3D bilinear interpolation and combining them with corresponding encoder outputs. A final $1 \times 1 \times 1$ convolution and sigmoid function produce the segmentation output. The VAE branch compresses the encoder output into a latent representation (mean and standard deviation), samples from it, and reconstructs the image using a decoder-like path without skip connections. The main features of SegResNet are using a VAE branch for better learning on small dataset sizes and using more blocks with residual connections in the encoder path compared to the decoder path and also using non-trainable operations in the upsampling process.

After the introduction of transformers and attention mechanisms and their success in language modeling and consequently using them in image processing tasks like vision transformers (ViTs), some researchers utilized them in medical image segmentation tasks. UNETR [4] follows a U-Net-like encoder-decoder structure, where the encoder is built entirely from transformer blocks operating on a sequence of non-overlapping 3D image patches. The input volume is divided into uniform patches, flattened, and linearly projected into a fixed K-dimensional embedding space, with positional embeddings added to retain spatial context. The transformer encoder comprises multiple layers of multi-head self-attention (MSA) and MLP blocks, using residual connections and layer normalization. Feature maps are extracted from different transformer layers (layers 3, 6, 9, 12), reshaped back into 3D tensors, and connected to the decoder through skip connections. The decoder progressively upsamples the features using deconvolutional layers, combines them with corresponding encoder outputs via concatenation, and applies $3 \times 3 \times 3$ convolutions and normalization. A final $1 \times 1 \times 1$ convolution with softmax activation produces the voxel-wise segmentation output. Swin UNETR [3] builds on the UNETR architecture by replacing the standard transformer encoder with a Swin Transformer [8], which introduces a more efficient way to model self-attention in 3D medical images. While UNETR processes the entire 3D volume as a sequence of fixed-size patches and applies global self-attention across all patches, Swin UNETR computes self-attention within local windows and shifts these windows between layers to allow communication between neighboring regions. This shifted window mechanism helps reduce computational cost while still capturing long-range dependencies. UNETR++ [12] builds on the UNETR architecture by introducing an efficient paired-attention (EPA) block to enhance feature representation. The EPA block combines spatial and channel attention using shared query-key pairs, enabling the model to efficiently capture both global spatial relationships and inter-channel dependencies. This dual-attention mechanism improves segmentation accuracy while maintaining low computational cost. Similar to the original U-Net, UNETR++ progressively reduces spatial resolution and increases the number of feature channels at each encoder stage.

MS-DSA-Net [15] also follows similar design principles to those used in the U-Net architecture [15], which remains foundational for medical image segmentation tasks. Its peculiarity is the integration of the parallel transformer pathways with dual self-attention (DSA) modules to enhance lesion segmentation.

Each DSA module combines spatial and channel self-attention using shared queries and keys but separate value paths, capturing both inter-position and inter-channel dependencies efficiently. Features are fused in the decoder through deconvolution and fusion blocks to recover spatial detail and generate precise probability maps. The reported results in [15] indicate that MS-DSA-Net shows the best performance among the existing architectures for the FCD detection task. Therefore, we adopt this architecture to apply the proposed TV loss.

3 Proposed Model

System architecture based on the MS-DSA-Net is given in Fig. 1.

The majority of studies, including the MS-DSA-Net, utilize *Dice Loss, Cross Entropy Loss,* or a combination of these two as the training loss. These loss functions are based on independent voxel prediction regarding the ground truth label of voxels. The main idea is to integrate a regularization term for output spatial smoothness in the loss function. This can teach the network to generate more consistent output maps so that the nearby voxels have similar probability values. To this end, we introduce an anisotropic Total Variation (TV) loss function.

The Total Variation loss is defined over the predicted voxel values to encourage spatial smoothness in the segmentation output.

Let $p_{i,j,k}$ be the predicted probability at voxel location (i, j, k), the isotropic TV loss $\mathcal{L}_{\text{TV}}$ is formally defined as:

$$\mathcal{L}_{\text{TV}} = \sum_{i,j,k} (|p_{i+1,j,k} - p_{i,j,k}| + |p_{i,j+1,k} - p_{i,j,k}| + |p_{i,j,k+1} - p_{i,j,k}|) \quad (1)$$

TV loss is widely used in image denoising and super-resolution models due to its ability to suppress noise while preserving edges. To the best of our knowledge, no previous study has incorporated TV loss into a UNet-based architecture for volumetric medical image segmentation, such as the detection of FCD in 3D MRI. It should be mentioned that the idea has been introduced for some 2D image segmentation tasks, such as the study performed by Javanmardi et al. [6]. In this paper, we integrate this regularization directly into the training loss function to promote contiguous and anatomically plausible segmentation in 3D space.

We adopt MS-DSA-Net [15] as the base architecture, as it achieves superior results on the FCD segmentation task among state-of-the-art methods. Our network consists of six encoder blocks, starting with 16 channels and doubling the number of channels at each stage until reaching 512 channels at the bottleneck. Correspondingly, the spatial resolution is halved at each stage via 2×2 max-pooling. Each encoder block includes a residual unit composed of two convolutional layers with instance normalization and leaky ReLU activation. The output of the two convolutions is added to the input through a residual connection, followed by another convolution and normalization layer.

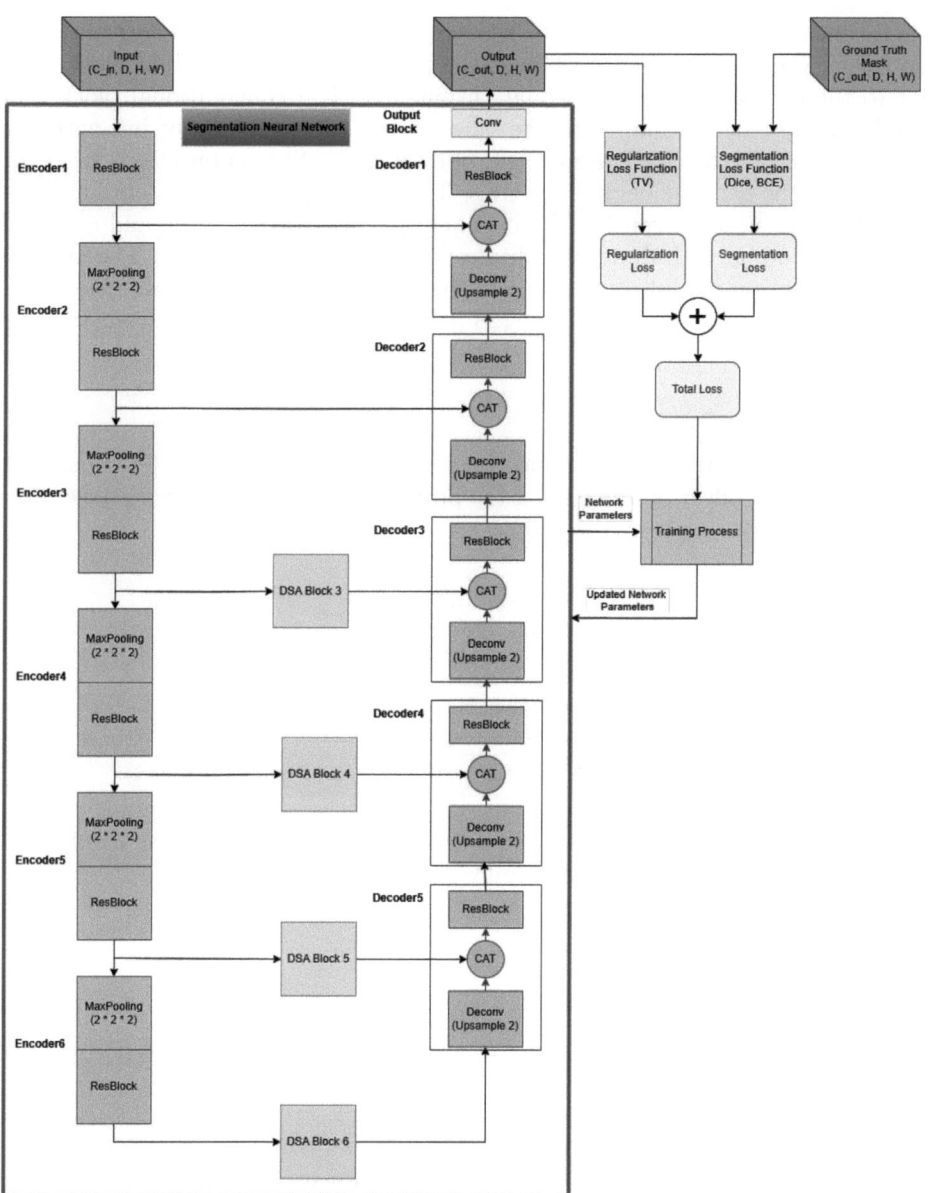

Fig. 1. The architecture of the system for FCD detection. It is based on the MS-DSA-Net (inside the red box) and the proposed TV loss function. (Color figure online)

The decoder path consists of five decoder blocks, each performing the inverse operations of its corresponding encoder: halving the number of channels and doubling the spatial resolution. Each decoder block starts with a deconvolution layer, then concatenates its output with the skip connection from the corresponding encoder (when available), and passes the result through a residual block similar to the encoder design. The final output layer is a convolutional block that produces a prediction map with the same spatial size as the input but with two channels: one for the background and one for FCD. Skip connections from the encoder to the decoder begin from stage 3 and employ dual self-attention transformers. These blocks consist of a channel-wise attention module and a spatial attention module with dimensionality reduction via a linear layer. The outputs of both attention modules are added to the input using residual connections. We used an input patch size of $128 \times 128 \times 128$, randomly selected from training subjects.

3.1 Loss Functions

The base loss for training was Dice Loss computed only on the FCD channel. To evaluate the effect of integrating the proposed regularization, we utilize three loss formulations, explained as follows:

Dice Loss [15]:
$$\mathcal{L}_{\text{Dice}} = 1 - \frac{2\sum_i p_i g_i + \epsilon}{\sum_i p_i + \sum_i g_i + \epsilon} \quad (2)$$

Binary Cross Entropy (BCE) Loss:
$$\mathcal{L}_{\text{BCE}} = -\frac{1}{N} \sum_{i=1}^{N} [g_i \log(p_i) + (1 - g_i) \log(1 - p_i)] \quad (3)$$

where p_i is the predicted probability, g_i is the ground truth label for voxel i, and $\epsilon = 1 \times 10^{-5}$ ensures numerical stability.

Total Loss:
- Dice + BCE (equal weight):
$$\mathcal{L}_{\text{Total}} = 0.5 \cdot \mathcal{L}_{\text{Dice}} + 0.5 \cdot \mathcal{L}_{\text{BCE}} \quad (4)$$
- Dice + TV (TV weight = 0.1):
$$\mathcal{L}_{\text{Total}} = 1.0 \cdot \mathcal{L}_{\text{Dice}} + 0.1 \cdot \mathcal{L}_{\text{TV}} \quad (5)$$

The weight of 0.1 is chosen for the TV loss because higher values may encourage trivial zero outputs, while lower values reduce its regularization impact (based on practical results). The TV loss is computed across three directions (x, y, z) without averaging, making the effective regularization equivalent to weighting the average by 0.3. The coefficient 1.0 for the Dice loss ensures that it remains the primary component guiding the segmentation. The TV term is weighted at 0.1 to act as a regularizer. Although the total sum exceeds 1.0, the loss terms are on different scales, and this combination was selected based on empirical performance. A lower weight for the TV term prevents over-smoothing or trivial solutions (e.g., empty masks), while still encouraging spatial consistency.

4 Experiments

4.1 Dataset and Preprocessing

We use the same dataset as in the MS-DSA-Net [15]. The dataset is available publicly [11]. It consists of T1 and FLAIR MRI modalities from 85 epilepsy patients and 85 healthy controls. For this study, only patient data was used and split randomly into training, validation, and test sets. Preprocessing (reorientation, skull stripping, modality alignment, and registration to MNI152 template space) was performed using the FSL toolkit[1].

4.2 Training

The training procedure involves the following steps:

- Random patch sampling with balanced FCD/background samples
- Data augmentation including random crop, rotation, flipping, intensity shift, and adding Gaussian noise
- Input patches fed as batches into the network
- Loss computation and backpropagation using AdamW optimizer
- Early stopping based on validation loss stagnation (patience threshold)

Weights were initialized using Kaiming normal for convolutional layers, Xavier uniform for linear and attention layers, and constants for normalization layers. A learning rate scheduler was used, beginning at 10% of the maximum rate, linearly warming up for 10 epochs, and followed by cosine annealing decay. Early stopping was employed by monitoring the validation loss. Training was halted if the loss did not improve for a specified number of consecutive epochs (a patience threshold of 25). All experiments were implemented using PyTorch and MONAI, incorporating code from the MS-DSA-Net repository[2]. We used the following configurations for training and evaluation:

- **Subject Split:** 59 training, 12 validation, and 14 test subjects
- **Patches per Image:** 4
- **Initial Learning Rate:** 1×10^{-4}
- **Minimum Learning Rate:** 1×10^{-6}
- **Max Epochs:** 300
- **Batch Size:** 1 (i.e., 4 patches of one subject per batch)
- **Early Stopping Patience:** 25 epochs
- **Total Trainable Parameters:** 43,524,802
- **Hardware:** NVIDIA GeForce RTX 2080 Ti (12 GB RAM)

---

[1] https://fsl.fmrib.ox.ac.uk/fsl/docs/.
[2] https://github.com/zhangxd0530/MS-DSA-Net.

4.3 Evaluation Metrics

Voxel-level validation metrics after each epoch included sensitivity (Sens), precision (Prec), and mean Dice score (DC). On the test set, we also computed i) subject-level sensitivity (sSens): presence of any true positive voxel match and ii) False Positive Clusters (nFPC): average number of falsely detected voxel clusters per subject.

4.4 Post-processing

After prediction, we applied connected component analysis with the following steps:

- Binary opening: dilation followed by erosion
- Binary hole filling with $5 \times 5 \times 5$ kernel
- Connected component labeling with 26-connectivity ($3 \times 3 \times 3$ structure)
- Cluster size filtering: removal of clusters smaller than 50 voxels

These post-processing steps and subject-level metrics are inspired by the base method, although specific hyperparameters were not disclosed. To account for randomness, each experiment was performed 10 times, and the reported values include the mean and standard deviation of the evaluated metrics. The same train, validation, and test splits were used across all scenarios.

4.5 Quantitative Results

As shown in Table 1, adding TV loss to the Dice loss leads to a noticeable improvement in the Dice score, both with and without post-processing. While the addition of BCE loss also shows gains, its impact is smaller compared to TV loss. Regarding the average number of false positive clusters (nFPC), TV loss consistently reduces this metric more effectively than BCE loss. Although BCE slightly improves voxel- and subject-level sensitivity more than TV, the precision is better when TV loss is used. Comparing pre- and post-processed results, it is evident that post-processing improves Dice score, precision, and nFPC across all loss functions, albeit with a slight reduction in sensitivity. Notably, the gain from post-processing with plain Dice loss (from 0.2811 to 0.2866) is smaller than the gain achieved by adding TV loss to Dice (from 0.2811 to 0.3104). Additionally, the original nFPC value with Dice+TV is already significantly better than with Dice alone, indicating stronger inherent regularization.

4.6 Qualitative Results

We visualize the segmentation results on a representative test subject to illustrate the effect of post-processing and the contribution of including TV loss during training (Fig. 2 and Fig. 3). Visualizations were generated using the MITK software[3]. Figure 2 illustrates the segmentation results produced by the base

---

[3] https://github.com/MITK/MITK.

model trained only with Dice Loss, before and after applying post-processing. In Fig. 2a, we can see a false positive cluster in the predicted mask (blue) that does not overlap with the ground truth (green), indicating a lack of spatial consistency in the raw network output. In Fig. 2b, after post-processing, the false positive cluster is successfully removed (yellow mask), confirming that connected component analysis can enforce smoother predictions as a post hoc fix.

Table 1. Evaluation results (mean ± standard deviation) on the test and post-processed test datasets: sensitivity (Sens), precision (Prec), mean Dice score (DC), subject-level sensitivity (sSens), and False Positive Clusters (nFPC).

(a) Test Results

Loss	Sens	Prec	DC	sSens	nFPC
Dice	0.3822 ± 0.0337	0.2767 ± 0.0417	0.2811 ± 0.0200	0.7857 ± 0.0337	22.0071 ± 7.5851
Dice + BCE	0.3964 ± 0.0439	0.2648 ± 0.0824	0.2885 ± 0.0473	0.8071 ± 0.0345	9.8071 ± 6.4381
Dice + TV	0.3845 ± 0.0425	0.3000 ± 0.0428	0.3104 ± 0.0246	0.8000 ± 0.0301	8.4500 ± 2.2591

(b) Post-processed Test Results

Loss	Sens	Prec	DC	sSens	nFPC
Dice	0.3765 ± 0.0333	0.2880 ± 0.0440	0.2866 ± 0.0204	0.7714 ± 0.0452	3.8786 ± 1.0118
Dice + BCE	0.3916 ± 0.0437	0.2735 ± 0.0858	0.2925 ± 0.0471	0.8071 ± 0.0345	3.1714 ± 1.0834
Dice + TV	0.3788 ± 0.0426	0.3102 ± 0.0444	0.3146 ± 0.0246	0.7928 ± 0.0226	3.1643 ± 0.7439

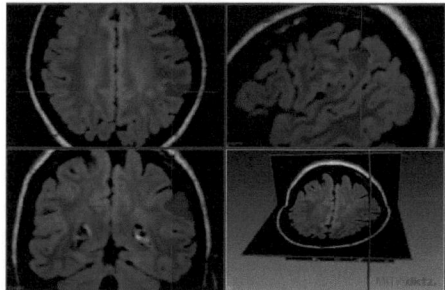

(a) Results of the base model. Green: ground truth mask; blue: predicted mask. Note the false positive cluster in the axial view (top-right pane).

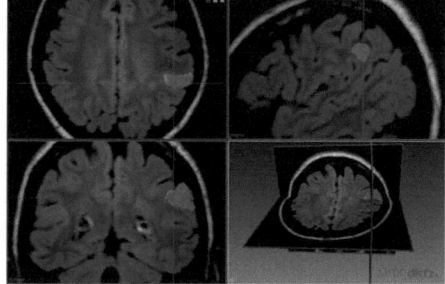

(b) Results after applying post-processing. Green: ground truth mask; yellow: predicted mask. Note that the false detection is removed.

Fig. 2. Comparison of predicted segmentation before and after post-processing for the base model. (Color figure online)

Figure 3 illustrates the segmentation results when the model is trained with the proposed TV loss added to Dice loss. In Fig. 3a, the predicted mask (blue) shows a high degree of overlap with the ground truth (green) and no visible

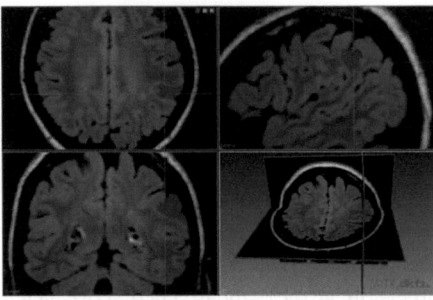 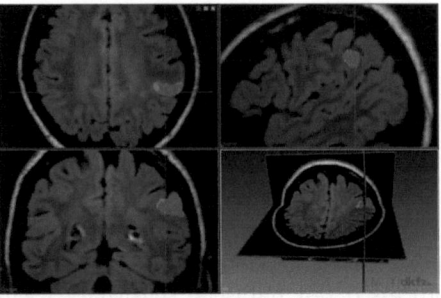

(a) Model trained with TV loss. Green: ground truth mask; blue: predicted mask. The false positives are no longer present.

(b) Model trained with TV loss after post-processing. Green: ground truth mask; yellow: predicted mask. Minimal change, indicating TV loss already enforced spatial consistency.

Fig. 3. Segmentation results with the proposed TV loss, before and after post-processing. (Color figure online)

false positives, even without post-processing. This highlights the regularizing effect of TV loss in enforcing spatial smoothness during training. In Fig. 3b, the result after post-processing (yellow mask) is nearly identical to the unprocessed prediction, confirming that TV loss had already smoothed the prediction to the extent that additional post-processing has minimal impact.

4.7 Discussion

Spatial consistency and smoothness in the prediction maps of a volumetric medical image segmentation network are desirable features that can be achieved by applying connected component analysis as a post-processing step on the results or can be seen as a constraint that can guide the network to learn features in a way that creates consistent and smooth outputs. Comparing the results on a base network that has the best results on the FCD detection task on MRI images in our experiments and adding the TV loss during the training process showed that its improvement to the test metrics is better than applying post-processing. Furthermore, applying post-processing to a model already trained with TV loss yields only a minimal additional improvement effect compared to the base one. Therefore, since the smoothness constraint has already been effectively incorporated during training, leaving little room for further enhancement through post-processing.

Dice Loss focuses mostly on the intersection between prediction and ground truth mask, while BCE loss encourages the voxel values to be close to 0 or 1 because the ground truth labels are either 0 or 1 for each voxel and the ground truth is essentially smooth and consistent, so at the voxel level BCE loss can help remove small false positive clusters and smooth predictions to an extent, but using TV loss encourages the network more to have smooth transitions of predictions between adjacent voxels. However, using TV loss could encounter a drawback because it can encourage the network to create an all-zero or all-one output map. This is a trivial solution that has TV loss = 0, and it should be handled by proper weighting of TV loss when it sums up to the original loss. Another drawback could be the removal of potentially small true positive regions, because, especially in FCD segmentation, having very small positive regions can be a case. It is also worth noting that identifying patients who are harder to treat is a common problem in medical research. Just as some epilepsy patients with FCD are difficult to diagnose and manage, other conditions—such as cardiac patients with allergies to donor organs—face similar challenges. These issues highlight the importance of improving segmentation methods for use in more complex clinical cases [5].

5 Conclusions

This paper introduced a Total Variation (TV) regularized framework for segmenting Focal Cortical Dysplasia (FCD) in three-dimensional (3D) brain MRI data. Our objective was to tackle a key issue in volumetric medical image segmentation: guaranteeing spatial consistency and anatomical plausibility in the anticipated results. We incorporated a smoothness requirement into the training process by enhancing a state-of-the-art transformer architecture (MS-DSA-Net) with an anisotropic TV loss term. Our experimental findings indicate that our straightforward yet efficient regularization approach surpasses conventional post-processing techniques, improving both voxel-level precision and overall segmentation consistency. Notably, we found that models trained using TV loss demonstrated remarkable internal consistency, rendering extra post-processing mostly superfluous—underscoring the efficacy of learning-based regularization. The proposed method enhances the current initiative to develop resilient and interpretable deep learning systems for clinical neuroimaging applications. Although our method is designed for detecting FCD in brain MRI images, the idea of adding Total Variation regularization can also be useful in other medical imaging tasks. For example, similar challenges exist in detecting small lung nodules or subtle cardiac scars in MRI scans. These tasks also require smooth and spatially coherent segmentations, which our approach supports. Future research can explore how this method performs in such diverse medical imaging problems. This methodology establishes a basis for further investigation of integrated regularization techniques, particularly in scenarios with limited training data and reduced lesion visibility. Future research may explore adaptive or region-specific regularization, incorporation of uncertainty estimation, or extensive evaluation on larger datasets.

Acknowledgments. We acknowledge the support of the PNRR project FAIR - Future AI Research (PE00000013), Spoke 9 - Green-aware AI, under the NRRP MUR program funded by the NextGenerationEU.

Disclosure of Interests. The authors have no competing interests to declare that are relevant to the content of this article.

References

1. Feigin, V.L., et al.: Global, regional, and national burden of epilepsy, 1990–2021: a systematic analysis for the global burden of disease study 2021. Lancet Publ. Health **10**(3), e203–e227 (2025)
2. Fisher, R.S., et al.: Ilae official report: a practical clinical definition of epilepsy. Epilepsia **55**(4), 475–482 (2014)
3. Hatamizadeh, A., Nath, V., Tang, Y., Yang, D., Roth, H.R., Xu, D.: Swin unetr: swin transformers for semantic segmentation of brain tumors in MRI images. In: Crimi, A., Bakas, S. (eds.) BrainLes 2021. LNCS, vol. 12962, pp. 272–284. Springer, Cham (2021). https://doi.org/10.1007/978-3-031-08999-2_22
4. Hatamizadeh, A., et al.: Unetr: transformers for 3d medical image segmentation. In: Proceedings of the IEEE/CVF Winter Conference on Applications of Computer Vision, pp. 574–584 (2022)
5. Haynatzki, R., et al.: Building artificial intelligence, machine learning, and causal models to improve cardiac health. J. Phys. Conf. Ser. **2910**, 012016. IOP Publishing (2024)
6. Javanmardi, M., Sajjadi, M., Liu, T., Tasdizen, T.: Unsupervised total variation loss for semi-supervised deep learning of semantic segmentation. arXiv preprint arXiv:1605.01368 (2016)
7. Jiménez-Murillo, D., et al.: Automatic detection of focal cortical dysplasia using MRI: a systematic review. Sensors **23**(16), 7072 (2023)
8. Liu, Z., et al.: Swin transformer: hierarchical vision transformer using shifted windows. In: Proceedings of the IEEE/CVF International Conference on Computer Vision, pp. 10012–10022 (2021)
9. Myronenko, A.: 3D MRI brain tumor segmentation using autoencoder regularization. In: Crimi, A., Bakas, S., Kuijf, H., Keyvan, F., Reyes, M., van Walsum, T. (eds.) BrainLes 2018. LNCS, vol. 11384, pp. 311–320. Springer, Cham (2019). https://doi.org/10.1007/978-3-030-11726-9_28
10. Ronneberger, O., Fischer, P., Brox, T.: U-net: convolutional networks for biomedical image segmentation. In: Navab, N., Hornegger, J., Wells, W.M., Frangi, A.F. (eds.) MICCAI 2015, Part III. LNCS, vol. 9351, pp. 234–241. Springer, Cham (2015). https://doi.org/10.1007/978-3-319-24574-4_28
11. Schuch, F., et al.: An open presurgery MRI dataset of people with epilepsy and focal cortical dysplasia type ii. Sci. Data **10**(1), 475 (2023)
12. Shaker, A., Maaz, M., Rasheed, H., Khan, S., Yang, M.H., Khan, F.S.: Unetr++: delving into efficient and accurate 3d medical image segmentation. IEEE Trans. Med. Imaging **43**(9), 3377–3390 (2024)
13. Splitkova, B., et al.: A new perspective on drug-resistant epilepsy in children with focal cortical dysplasia type 1: from challenge to favorable outcome. Epilepsia **66**(3), 632–647 (2025)
14. Symms, M., Jäger, H., Schmierer, K., Yousry, T.: A review of structural magnetic resonance neuroimaging. J. Neurol. Neurosurg. Psychiatry **75**(9), 1235–1244 (2004)
15. Zhang, X., et al.: Focal cortical dysplasia lesion segmentation using multiscale transformer. Insights Imaging **15**(1), 222 (2024)

Enhancing Flight Delay Prediction with Network-Aware Ensemble Learning

Mary Dufie Afrane, Yao Xu(✉), and Lixin Li

Georgia Southern University, Statesboro, GA, USA
{ma20542,yxu,lli}@georgiasouthern.edu

Abstract. This study presents a comprehensive framework for predicting departure delays in U.S. domestic aviation by integrating advanced feature engineering, network analysis, and ensemble learning methods. Using a dataset of 2,638,673 flights across 354 airports from May to August 2024, we engineered predictors using temporal features (cyclical time), operational metrics (airport congestion), and network characteristics (in-/out-degree centrality and cluster labels). We extracted data for the five airlines with the highest number of flights: Southwest (WN), American (AA), Delta (DL), United (UA), and SkyWest (OO). A novel greedy mutual information and correlation-based feature selection method was then applied to each dataset to improve prediction performance. Multiple classifiers, including Random Forest (RF), Extra Trees (ET), XGBoost, and LightGBM, were evaluated. RF and ET consistently outperformed the others, motivating their inclusion in a Voting ensemble. The Voting classifier achieved robust performance across all five airlines, with overall accuracy ranging from 88.9% to 91.8%, F1–scores between 88.5% and 91.4%, and AUC–ROC values all above 95%. DL yielded the highest performance (91.8% accuracy and 96.8% AUC–ROC). These results demonstrate that combining network–cluster information with rich historical features substantially improves delay prediction, providing a scalable approach for airlines and air traffic managers to mitigate operational disruptions.

Keywords: Flight delay prediction · Network clustering · Feature selection · Ensemble learning · Voting classifier

1 Introduction

Flight delays represent a significant operational challenge for airlines, airports, and passengers, leading to increased costs, disrupted itineraries, and cascading network effects. In 2023, U.S. airlines reported billions of dollars in delay-related costs, highlighting the critical need for accurate delay forecasting tools [12]. Traditional statistical models have gradually given way to machine learning and deep learning approaches capable of capturing complex spatial, temporal, and network-level dependencies. Recent studies have employed random forest (RF) with engineered temporal and network features, gradient boosting (GB) algorithms optimized via hyperparameter tuning, and deep operator networks (DeepONet) integrated with metaheuristic searches to attain improvements in predictive accuracy and robustness. Despite these advances, many existing models suffer from limited generalizability due to geographically or temporally constrained datasets and often neglect the interplay between airport network topology and weather patterns. Moreover, little work has investigated airline-specific performance variation within network clusters.

To address these gaps, our study constructs a dataset using multi-source flight and weather data from summer 2024, applies a novel greedy mutual information (MI) and correlation-based feature selection method to eliminate redundant predictors, and leverages airport connectivity clustering to capture latent route structures. After evaluating several classifiers, including RF, Extra Trees (ET), XGBoost, and LightGBM, RF and ET consistently outperformed the others across all datasets, motivating their use in a Voting ensemble. By training the Voting classifier on both individual airlines and network clusters, we proposed a robust and scalable framework for operational departure delay prediction across diverse airline networks.

The remainder of the paper is organized as follows. Section 2 provides a detailed review of related literature. Section 3 describes the dataset, outlines the preprocessing and feature engineering steps, and presents the methodology. Section 4 presents the experimental results, highlighting model performance and key insights. Finally, Sect. 5 summarizes the main contributions and discusses directions for future research.

2 Literature Review

The prediction of flight delay has been extensively studied using a variety of methodologies [15]. The reviewed research emphasizes improving predictive accuracy and robustness by combining advanced algorithms, effective feature engineering, and comprehensive data processing. This review of the literature highlights recent contributions in the field, highlighting diverse approaches to enhance prediction capabilities.

Hybrid models combine multiple algorithms or techniques to leverage their strengths, leading to enhanced predictive performance. Dai [9] introduced a hybrid machine learning framework employing DBSCAN clustering, ANOVA, Forward Sequential Feature Selection (FSFS), and a Coyote Optimization Algorithm (COA)-optimized Weighted RF (COWRF) for predicting flight arrival delays. The dataset included 24,986 U.S. flight records from JFK Airport in select months of 2023, initially with 36 features narrowed down to 21 impactful predictors. The model attained an impressive accuracy of 97.2%, recall of 97.34%, precision of 88.13%, and an F-measure of 0.9193. Despite the outstanding results, the localized nature of the dataset potentially limits generalizability.

RF and GB models are widely recognized for their robust performance and adaptability to various types of structured data. Afrane et al. [1] proposed a robust RF classifier enhanced with domain-specific feature engineering, including rolling averages, cyclic temporal encodings, and network centrality measures.

They utilized 354,452 records of U.S. domestic flights from July 2024 combined with weather data, achieving 92.36% accuracy, 98.29% precision, and an F1-score of 91.86%. Similarly, Ajayi et al. [2] emphasized network centrality features (degree, betweenness, and closeness) in RF, GB, and CatBoost (CB) models, applied to over 6.9 million U.S. flights from July 2022 to June 2023. The RF model performed best, achieving 86.2% accuracy and an F1-score of 81.2, demonstrating the benefit of integrating network features into prediction models. Li et al. [25] proposed an RF model integrating prioritized weather and non-weather features based on clustering, achieving a notable improvement with 96% accuracy and 86% recall on U.S. flight data. Paramita et al. [28] employed RF in a PySpark-based cluster computing environment, attaining 92.7% accuracy while significantly reducing computation time using a publicly available flight dataset sourced from GitHub.

Comparative studies of GB algorithms highlight their effectiveness and emphasize the importance of algorithm selection and hyperparameter tuning. Alfarhood et al. [3] investigated multiple machine learning algorithms, including CB, XGBoost, LightGBM, RF, and Multi-layer Perceptron (MLP), for predicting flight delays using 775,000 Saudi Arabian domestic flights from 2017–2023. CB emerged superior with 76% accuracy for classification and a MAE of 12.19 min for regression tasks. Kiliç and Sallan [20] also evaluated GB Machine alongside RF and Logistic Regression, demonstrating superior performance with 75% accuracy, 88% recall, and an AUC-ROC (the area under the receiver operating characteristic curve) of 0.89 on 5.6 million U.S. flights. Khan et al. [18] compared GB algorithms (CB, LightGBM, XGBoost) using 539,383 U.S. flight records, identifying CB as most effective, achieving 68% accuracy and 69% after hyperparameter tuning. Similarly, Hatıpoğlu et al. [14] used GB models (XGBoost, LightGBM, CB) with Bayesian optimization on 18,148 flights from a Turkish airline, with LightGBM achieving the highest accuracy of 96.7%.

Deep learning approaches exploit complex neural architectures capable of handling large volumes of data and capturing intricate patterns for delay prediction. Bisandu and Moulitsas [5] combined DeepONet with the Gradient-Mayfly Optimization Algorithm (GMOA) using over 900,000 U.S. domestic flights, achieving an RMSE of 0.0765. Mamdouh et al. [26] utilized an Attention-Based Bidirectional LSTM (ATT-BI-LSTM) on 5.5 million U.S. flights enriched with weather data, achieving 94.3% testing accuracy. Falque et al. [11] employed LightGBM on 10 million records from Paris Charles de Gaulle Airport, achieving a remarkable MAE of 8.19 min. Khan et al. [19] introduced an Adaptive Bidirectional Extreme Learning Machine (AB-ELM), achieving an accuracy of 80.66% in predicting delay subcategories and durations on international flights from Hong Kong.

Spatio-temporal models capture the geographic and time-dependent dynamics of flight delays by integrating spatial correlations with temporal dependencies. Li and Revesz [22] introduced spatio-temporal interpolation techniques using shape functions, laying early groundwork for modeling high-dimensional temporal-spatial data. Building on this foundation, Anderson and Revesz [4] developed efficient operators for querying moving objects, while Revesz [30] provided a comprehensive framework for spatio-temporal data management, including applications in transportation systems. More recent work includes Shen et al. [32], who applied a hybrid federated deep learning model with a diffusion graph convolutional network and residual gated recurrent unit, handling over 12 million U.S. flight records with an R^2 of 0.8532 and RMSE of 6.9755 min. Li et al. [23] implemented a CNN-LSTM-RF hybrid model capturing spatial-temporal dependencies across 5.4 million U.S. flights, achieving an accuracy of 92.39%. Zheng et al. [35] utilized a Spatial-Temporal Gated Multi-Attention Graph Network (STGMAGNet) on 7.28 million U.S. flights, achieving a 21% reduction in MAE compared to competitors. Li and Jing [24] proposed a spatial-temporal RF model, achieving 92.39% accuracy on 762,415 Chinese domestic flights by integrating network and weather features.

Clustering and graph-based models leverage structural insights from airport networks to improve predictive accuracy through enhanced feature extraction and pattern recognition. Deng et al. [10] developed a Cluster Clustering-Based Modular Integrated Deep Neural Network (CC-MIDNN) using k-means clustering on 16,347 flights from Lisbon Airport, achieving a significant MAE reduction. Wei et al. [33] leveraged a PSO-K-means-based TS-BiLSTM-Attention model on nearly one million Chinese domestic flights, significantly outperforming single-airport models. Cai et al. [7] proposed a Geographical and Operational Graph Convolutional Network (GOGCN), achieving the lowest MAE (7.742 min) among competing models on 1.8 million flights from the Civil Aviation Administration of China.

Effective feature engineering and ensemble methods can significantly enhance predictive accuracy by identifying and utilizing the most relevant data attributes. Foundational work on constraint query languages by Kanellakis et al. [16,17] provides theoretical support for structured, constraint-based feature representations in predictive modeling. Pineda-Jaramillo et al. [29] integrated explainable AI techniques with Linear Discriminant Analysis, achieving a high recall of 0.824 using flight data from Santiago de Chile International Airport. Schösser and Schönberger [31] utilized XGBoost and highlighted the importance of short-term features like previous delays on U.S. flights, achieving 89.9% accuracy. Yi et al. [34] applied the Stacking ensemble method on nearly 300,000 flights from Logan International Airport in Boston, significantly improving performance stability with 82.2% accuracy for departure delays. Chakrabarty et al. [8] employed a GB Classifier and Recursive Feature Elimination to select influential features from over 1.2 million U.S. flight records, achieving 88.6% accuracy and 85.4% precision.

These studies collectively present a diverse range of methodologies, including traditional machine learning, advanced deep learning, and spatial-temporal techniques, significantly enhancing performance in flight delay prediction tasks. However, many approaches are constrained by dataset limitations, such as geographical specificity and restricted temporal scope, potentially affecting generalizability. Our study addresses these limitations by performing a comprehensive airline analysis, aimed at achieving broader applicability and robust predictive insights across varied aviation environments.

3 Methodology

The first step of developing the flight delay prediction model was thoroughly exploring and cleaning the raw data to ensure that only accurate, relevant, and high-quality records were used in the modeling pipeline. The dataset comprised 2,638,673 flights across 354 airports from the Bureau of Transportation Statistics (BTS) [6] and corresponding weather observations from Open-Meteo [27] covering May through August 2024. Table 1 summarizes the features extracted from the raw data and used in subsequent preprocessing and analysis.

Table 1. Original features in the dataset.

Feature Type	Description
Flight	Year, Month, Day of Month, Day of Week, Airline, Flight Number, Tail Number, Origin, Destination, Scheduled Departure Time, Scheduled Arrival Time, Departure Delay, Arrival Delay, Number of Seats
Weather	Temperature, Relative Humidity, Dewpoint, Surface Pressure, Precipitation, Rain, Snowfall, Cloudcover (Low, Mid, High), Wind Speed, Wind Direction, Wind Gusts

3.1 Data Cleaning and Preprocessing

To ensure data quality and improve model reliability, several filtering steps were applied during preprocessing. Flights that were cancelled or diverted were removed, as they do not contribute to meaningful delay prediction. Records with missing or null values in key delay-related columns (departure delay or arrival delay) were also excluded. To further reduce noise, outliers in departure and arrival delays were identified using z-scores; specifically, records with absolute z-scores above 3 were treated as outliers and removed. Additionally, flights that depart less than twice a week were removed. The target variable, departure delay, was created by applying a threshold of 15 min to the departure delay column. Flights departing 15 min or more after their scheduled time were labeled as delayed, and those departing sooner were labeled as on-time. This threshold aligns with the industry standard used by the U.S. Department of Transportation for classifying flight delays [13].

3.2 Feature Engineering

A rich set of predictive features was engineered to capture temporal, operational, spatial, and behavioral dimensions of each flight. Temporal features, including the day of the week and day of the year, were encoded using sine and cosine transformations to preserve their cyclical nature. Similarly, scheduled departure and arrival times were transformed into sinusoidal components to reflect their periodic behavior across a 24-h day. Operational traffic context was incorporated by calculating the number of flights scheduled to arrive at or depart from each airport on the same day. This congestion-related information helps capture the potential impact of airport traffic on delay likelihood.

To capture the structural importance of each airport within the national air traffic network, a directed graph was constructed where nodes represent airports and directed edges denote flight routes. Two graphs were built: a weighted version, where edge weights correspond to the total number of flights between origin-destination pairs, and an unweighted version capturing only the presence of routes. Network centrality metrics, including in-degree and out-degree centrality, were calculated for each airport to capture the flow of traffic entering and leaving the airport.

To provide temporal context and capture recent performance trends, historical delay statistics were calculated at multiple levels. For each flight number, rolling averages, standard deviations, and maximum values of both departure and arrival delays were computed over the past two weeks. At the tail-number level, historical delays due to late-arriving aircraft were calculated over the past three weeks to reflect aircraft-specific reliability. Additionally, 24-h delay summaries were generated for each origin and destination airport, including the average, standard deviation, and maximum delay values observed over the past 24 h.

3.3 Clustering Based on Airport Connectivity

To uncover latent structure in airport connectivity patterns, clustering was performed using k-means on the centrality features derived from the flight network. The elbow method was employed to determine the optimal number of clusters. This involved computing the within-cluster sum of squares (WCSS) for values of k ranging from 1 to 10 as shown in Fig. 1.

After selecting an appropriate k (in this case, 5), the k-means algorithm was applied to assign a cluster label to each origin-destination pair. These clusters effectively grouped routes with similar network characteristics. For example, a route connecting two highly connected hub airports might be assigned a different cluster than a route involving smaller regional airports.

Once cluster labels were assigned, they were merged back into the main flight dataset. This enriched dataset allowed models to consider cluster-level characteristics during training, potentially capturing how network topology influences delay patterns.

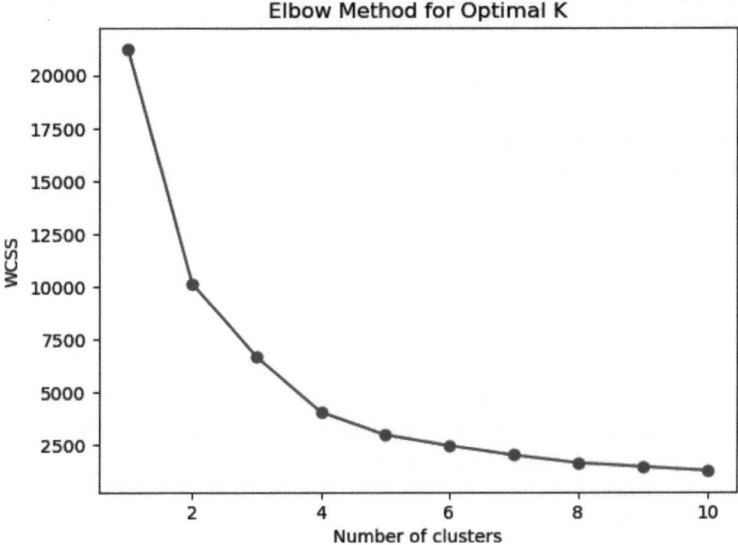

Fig. 1. The results of elbow method for optimal k.

3.4 Feature Selection

After clustering airport routes using network centrality metrics, the top five airlines by flight volume from May to August 2024 were extracted: Southwest (WN), American (AA), Delta (DL), United (UA), and SkyWest (OO). Each airline's flight data was then segmented by its corresponding cluster label computed in Sect. 3.3.

To identify the most informative predictors, mutual information (MI) was calculated between each feature and the target binary variable, the departure delay class. Features with near-zero MI scores (≤ 0.00001) were considered uninformative and removed from further analysis. To reduce multicollinearity, the remaining features were further refined using a greedy MI-correlation feature selection method, which iteratively selects features with the highest MI scores while discarding those that exhibit strong pairwise correlations (correlation $>$ 0.7) with already selected features. This approach prioritizes features with high MI scores and eliminates redundant ones based on pairwise correlations. The full procedure is detailed in Algorithm 1.

Algorithm 1. Greedy MI–Correlation Feature Selection

Input:
- *ordered_feats*: list of features sorted by descending MI scores
- *corr_matrix*: matrix of pairwise correlations
- *corr_thresh*: correlation threshold (default: 0.7)

Output: list of selected features with no pair having $|corr| > corr\_thresh$

1: *remaining* ← *ordered_feats*
2: *selected* ← [ ]
3: **for** *feat1* **in** *ordered_feats* **do**
4:    **if** *feat1* ∈ *remaining* **then**
5:       add *feat1* to *selected*
6:       *to_remove* ← { }
7:       **for** *feat2* **in** *ordered_feats* **do**
8:          **if** *feat2* ≠ *feat1* **then**
9:             **if** $|corr\_matrix[feat1][feat2]| > corr\_thresh$ **then**
10:                add *feat2* to *to_remove*
11:             **end if**
12:          **end if**
13:       **end for**
14:       *remaining* ← *remaining* \ *to_remove*
15:    **end if**
16: **end for**
17: **return** *selected*

3.5 Model Selection

We evaluated several classification models, including RF, ET, XGBoost, and LightGBM, across all airline datasets. Among these, RF and ET consistently demonstrated superior and stable predictive performance. This motivated their inclusion in a Voting ensemble to leverage their strengths and further improve generalization.

Finally, for each airline-cluster dataset, the following models were applied:

- **RF Classifier**: A bootstrap-aggregated ensemble of decision trees, effective in handling nonlinearities and interactions among features.
- **Extra Trees (ET) Classifier**: Similar to RF but introduces more randomness in the feature splitting process, helping reduce overfitting and computational cost.
- **Voting Classifier**: An ensemble meta-model that combines predictions from the RF and ET classifiers using soft voting (averaged probabilities), enhancing generalization and reducing variance.

Each model was trained using an 80/20 train–test split with stratified sampling to preserve the class distribution.

4 Results and Discussion

To evaluate the models' effectiveness across different clusters and airlines, we computed the accuracy, precision, recall, F1-score, and AUC-ROC on the test set.

The overall performance of the voting classifier models across the five major U.S. airlines (WN, AA, DL, UA, and OO) is presented in Table 2 and Fig. 2. These results represent average values computed across all five route-based clusters for each airline.

Table 2. Overall Performance of Voting Classifier by Airline

Airline	Accuracy	Precision	Recall	F1-Score	AUC-ROC
WN	0.889	0.914	0.859	0.885	0.960
AA	0.908	0.946	0.865	0.903	0.963
DL	**0.918**	0.958	**0.875**	**0.914**	**0.968**
UA	0.901	0.949	0.848	0.896	0.959
OO	0.913	**0.969**	0.854	0.908	0.955

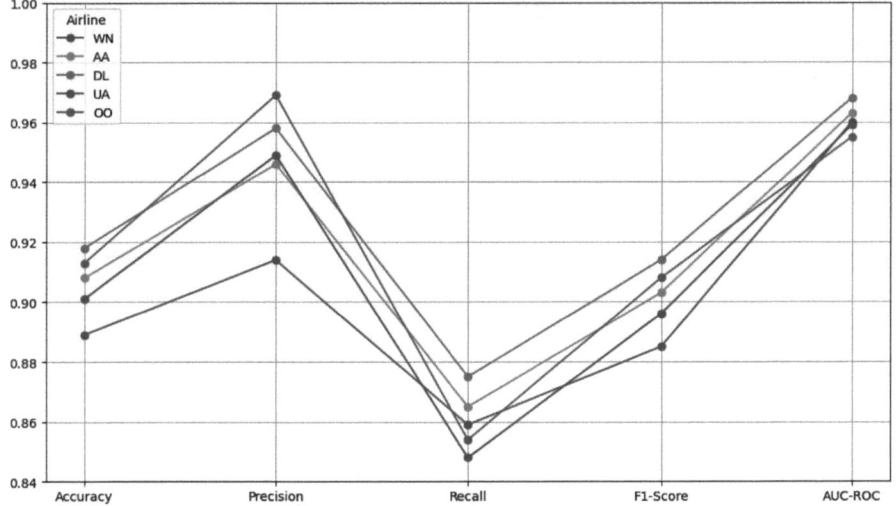

Fig. 2. Overall Performance Metrics by Airline

4.1 Overall Model Performance

All five airlines achieved strong classification performance, with accuracy ranging from 88.89% to 91.79%, precision from 91.4% to 96.9%, recall between 84.8% and 87.5%, F1–scores between 88.5% and 91.4%, and AUC-ROC scores consistently exceeding 95%. These results demonstrate the effectiveness of the Voting Classifier, which integrates predictions from the RF and ET models. The robust performance can be attributed to the inclusion of network-based clustering information as well as detailed historical and environmental features.

4.2 Top Performing Airlines

Among the five airlines, DL achieved the highest accuracy (91.79%), recall (87.46%), and F1-score (91.40%), demonstrating the most balanced and effective predictive capability. This suggests a strong alignment between model predictions and actual delay events across DL's network. Notably, OO achieved the highest precision (96.91%), indicating high confidence in delay predictions when they occurred, albeit with a trade-off in recall. AA and UA also delivered reliable results, with F1-scores of 90.31% and 89.56%, respectively. Their precision-recall balance was favorable, making them consistent across all metrics. WN had the lowest accuracy (88.89%) and recall (85.94%) among the five airlines. While still achieving an F1-score of 88.55% and AUC-ROC of 95.95%, this slightly reduced performance may stem from more complex or regionally diverse route networks.

4.3 Comparison with Previous Works

To contextualize the effectiveness of our approach, we compare the best-performing model in our study (DL) and the average performance across all five airlines against several prior departure delay prediction studies that used the same data source and evaluation metrics. Detailed comparisons are presented in Table 3 and Fig. 3.

Table 3. Comparison with previous works.

Model	Accuracy (%)	Precision (%)	Recall (%)	F1-score (%)
DL	91.8	95.8	87.5	91.4
Avg of the 5 Airlines	90.6	94.7	86.0	90.1
Kim and Park [21]	85.2	83.5	82.6	85.6
Khan et al. [18]	68.0	65.0	50.0	57.0
Yi et al. [34]	82.2	82.9	81.2	82.1

Our DL-specific model achieved an accuracy of 91.8%, precision of 95.8%, recall of 87.5%, and an F1-score of 91.4%, surpassing the performance reported in the previous three studies. Additionally, the average performance across our five airline models (accuracy of 90.6%, precision of 94.7%, recall of 86.0%, and F1-score of 90.1%) also exceeds prior results.

These comparisons confirm that our methodology offers strong predictive performance and generalization across diverse airline operations, positioning it well within current flight delay prediction research.

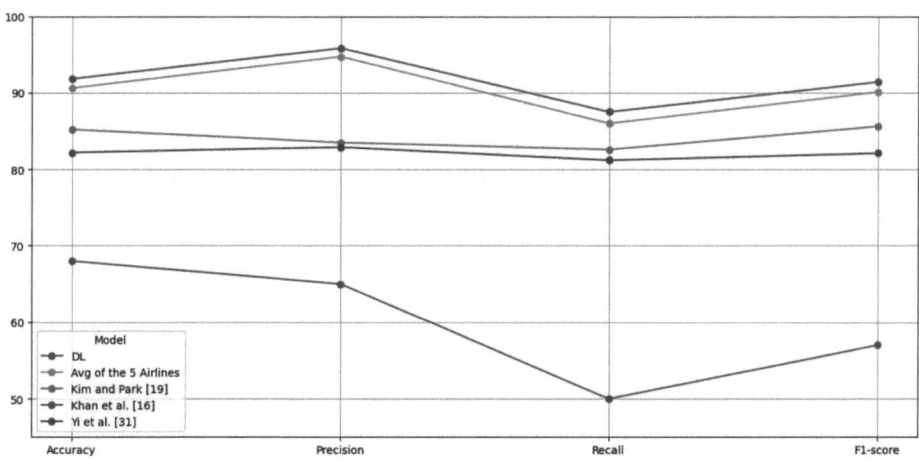

Fig. 3. Comparison with previous works.

5 Conclusion

This study demonstrates that integrating rich temporal, operational, and network-based features with ensemble learning yields highly accurate and reliable flight delay predictions. The proposed Voting Classifier, which combines RF and ET models, consistently outperformed single-model baselines and other classifiers such as XGBoost and LightGBM across all five major U.S. airlines. It achieved accuracy up to 91.8%, precision up to 96.9%, recall up to 87.46%, F1-score up to 91.40%, and AUC-ROC values consistently exceeding 95%. Our key methodological contributions include:

– Incorporating airport connectivity clustering to capture route-level heterogeneity,
– Engineering features that reflect recent delay patterns across multiple time scales,
– Proposing and applying a systematic MI-correlation-based feature selection process to reduce multicollinearity, and
– Segmenting data by airline–cluster combinations to better tailor models to operational characteristics.

By offering a scalable and interpretable prediction pipeline, this study provides practical insights for airlines and air traffic authorities aiming to proactively manage delay risks and improve passenger experience.

While the results confirm the effectiveness of our approach, certain limitations remain. The models were trained on summer 2024 flight data and may require retraining or adaptation for other seasons to maintain performance. Future work will focus on integrating the framework with real-time operational data, generalizing the pipeline for use with other datasets, applying transfer learning techniques for low-volume airlines, and extending the methodology to predict delay durations in addition to binary classifications.

References

1. Afrane, M.D., Xu, Y., Li, L., Wang, K.: Flight delay prediction using random forest with enhanced feature engineering. In: SoutheastCon 2025, pp. 1055–1056. IEEE (2025)
2. Ajayi, J., Xu, Y., Li, L., Wang, K.: Enhancing flight delay predictions using network centrality measures. Information **15**(9), 559(1–11) (2024)
3. Alfarhood, M., Alotaibi, R., Abdulrahim, B., Einieh, A., Almousa, M., Alkhanifer, A.: Predicting flight delays with machine learning: A case study from Saudi Arabian airlines. Int. J. Aerosp. Eng. **2024**(1), 3385463 (2024)
4. Anderson, S., Revesz, P.: Efficient maxcount and threshold operators of moving objects. GeoInformatica **13**(4), 355–396 (2009)
5. Bisandu, D.B., Moulitsas, I.: Prediction of flight delay using deep operator network with gradient-mayfly optimisation algorithm. Expert Syst. Appl. **247**, 123306 (2024)
6. Bureau of Transportation Statistics: On-Time Performance Data (2024). https://www.transtats.bts.gov/ONTIME/. Accessed 29 Apr 2025
7. Cai, K., Li, Y., Zhu, Y., Fang, Q., Yang, Y., Du, W.: A geographical and operational deep graph convolutional approach for flight delay prediction. Chin. J. Aeronaut. **36**(3), 357–367 (2023)
8. Chakrabarty, N., Kundu, T., Dandapat, S., Sarkar, A., Kole, D.K.: Flight arrival delay prediction using gradient boosting classifier. In: Abraham, A., Dutta, P., Mandal, J., Bhattacharya, A., Dutta, S. (eds.) Emerging Technologies in Data Mining and Information Security. Advances in Intelligent Systems and Computing, vol. 813, pp. 651–659. Springer, Singapore (2019). https://doi.org/10.1007/978-981-13-1498-8_57
9. Dai, M.: A hybrid machine learning-based model for predicting flight delay through aviation big data. Sci. Rep. **14**(1), 4603 (2024)
10. Deng, W., Li, K., Zhao, H.: A flight arrival time prediction method based on cluster clustering-based modular with deep neural network. IEEE Trans. Intell. Transp. Syst. (2023)
11. Falque, T., Mazure, B., Tabia, K.: Machine learning for predicting off-block delays: a case study at Paris–Charles de Gaulle international airport. Data Knowl. Eng. **152**, 102303 (2024)
12. Federal Aviation Administration: Air traffic by the numbers (2023). https://www.faa.gov/air_traffic/by_the_numbers. Accessed 12 May 2025
13. Federal Aviation Administration: Types of Delay (2024). https://aspm.faa.gov/aspmhelp/index/Types_of_Delay.html. Accessed 29 Apr 2025
14. Hatıpoğlu, I., Tosun, Ö., Tosun, N.: Flight delay prediction based with machine learning. LogForum **18**(1) (2022)
15. Huynh, T.K., Cheung, T., Chua, C.: A systematic review of flight delay forecasting models. In: 2024 7th International Conference on Green Technology and Sustainable Development (GTSD), pp. 533–540. IEEE (2024)
16. Kanellakis, P.C., Kuper, G.M., Revesz, P.Z.: Constraint query languages (preliminary report). In: Proceedings of the Ninth ACM SIGACT-SIGMOD-SIGART Symposium on Principles of Database Systems. PODS '90, pp. 299–313. Association for Computing Machinery, New York, NY, USA (1990)
17. Kanellakis, P.C., Kuper, G.M., Revesz, P.Z.: Constraint query languages. J. Comput. Syst. Sci. **51**(1), 26–52 (1995)

18. Khan, R., Akbar, S., Zahed, T.A.: Flight delay prediction based on gradient boosting ensemble techniques. In: 2022 16th International Conference on Open Source Systems and Technologies (ICOSST), pp. 1–5. IEEE (2022)
19. Khan, W.A., Chung, S.H., Eltoukhy, A.E., Khurshid, F.: A novel parallel series data-driven model for iata-coded flight delays prediction and features analysis. J. Air Transp. Manag. **114**, 102488 (2024)
20. Kiliç, K., Sallan, J.M.: Study of delay prediction in the us airport network. Aerospace **10**(4), 342 (2023)
21. Kim, S., Park, E.: Prediction of flight departure delays caused by weather conditions adopting data-driven approaches. J. Big Data **11**(1), 11 (2024)
22. Li, L., Revesz, P.: Interpolation methods for spatio-temporal geographic data. Comput. Environ. Urban Syst. **28**(3), 201–227 (2004)
23. Li, Q., Guan, X., Liu, J.: A CNN-LSTM framework for flight delay prediction. Expert Syst. Appl. **227**, 120287 (2023)
24. Li, Q., Jing, R.: Flight delay prediction from spatial and temporal perspective. Expert Syst. Appl. **205**, 117662 (2022)
25. Li, Q., Jing, R., Dong, Z.S.: Flight delay prediction with priority information of weather and non-weather features. IEEE Trans. Intell. Transp. Syst. **24**(7), 7149–7165 (2023)
26. Mamdouh, M., Ezzat, M., Hefny, H.: Improving flight delays prediction by developing attention-based bidirectional LSTM network. Expert Syst. Appl. **238**, 121747 (2024)
27. Open-Meteo: Free Weather API (2024). https://open-meteo.com/. Accessed 29 Apr 2025
28. Paramita, C., Supriyanto, C., Syarifuddin, L.A., Rafrastara, F.A.: The use of cluster computing and random forest algorithm for flight delay prediction. Int. J. Comput. Sci. Inf. Secur. (IJCSIS) **20**(2) (2022)
29. Pineda-Jaramillo, J., Munoz, C., Mesa-Arango, R., Gonzalez-Calderon, C., Lange, A.: Integrating multiple data sources for improved flight delay prediction using explainable machine learning. Res. Transp. Bus. Manag. **56**, 101161 (2024)
30. Revesz, P.Z.: Introduction to Databases: From Biological to Spatio-temporal. Springer, London, UK (2010). https://doi.org/10.1007/978-1-84996-095-3
31. Schösser, D., Schönberger, J.: On the performance of machine learning based flight delay prediction–investigating the impact of short-term features. Promet-Traffic&Transportation **34**(6), 825–838 (2022)
32. Shen, X., Chen, J., Yan, R.: A spatial-temporal model for network-wide flight delay prediction based on federated learning. Appl. Soft Comput. **154**, 111380 (2024)
33. Wei, X., Li, Y., Shang, R., Ruan, C., Xing, J.: Airport cluster delay prediction based on TS-bilSTM-attention. Aerospace **10**(7), 580 (2023)
34. Yi, J., Zhang, H., Liu, H., Zhong, G., Li, G.: Flight delay classification prediction based on stacking algorithm. J. Adv. Transp. **2021**(1), 4292778 (2021)
35. Zheng, H., et al.: A graph multi-attention network for predicting airport delays. Transp. Res. Part E: Logist. Transp. Rev. **181**, 103375 (2024)

Distributed Systems

FedMod: Vertical Federated Learning Using Multi-server Secret Sharing

Kasra Mojallal[✉], Ali Abbasi Tadi, and Dima Alhadidi

University of Windsor, 401 Sunset Ave, Windsor, ON N9B 3P4, Canada
{mojalla,abbasit,dima.alhadidi}@uwindsor.ca

Abstract. Vertical Federated Learning (VFL) allows multiple entities with feature-partitioned datasets to collaboratively train machine learning models while keeping their data private. However, many existing VFL approaches either rely on a single trusted server or incur significant computational and communication overhead due to encryption-based techniques. In this paper, we propose FedMod, a scalable and lightweight VFL framework that removes the need for trusted parties or cryptographic primitives. FedMod introduces a novel multi-server architecture combined with additive secret sharing to protect intermediate computations during training. We conduct extensive experiments across multiple real-world datasets and benchmark FedMod against state-of-the-art methods including homomorphic encryption, differential privacy, and functional encryption. Our results show that FedMod achieves comparable or superior accuracy with significantly lower computation time and communication cost. Moreover, FedMod provides strong protection in the semi-honest setting and remains secure even when some parties or servers partially collude. These results highlight FedMod's practicality for real-world privacy-preserving collaborative learning scenarios.

Keywords: privacy-preserving machine learning · vertical federated learning · secret sharing · distributed machine learning

1 Introduction

AI and machine learning are transforming fields such as healthcare and cybersecurity by enabling predictive tools like cancer diagnostics and threat detection systems. These capabilities rely on large volumes of sensitive data. However, regulations such as HIPAA [12] and PIPEDA [20] restrict centralized data sharing to protect individual privacy. Federated Learning (FL) offers a solution by training models across distributed data sources without moving data. FL is categorized into horizontal and vertical types. VFL is applicable when parties share sample identifiers but hold different features. For example, hospitals can collaboratively learn from patient data while retaining control over distinct attributes. Despite its advantages, VFL exposes intermediate results to inference attacks.

To address this, recent solutions apply Homomorphic Encryption (HE) [10] or Functional Encryption (FE) [3], which offer strong privacy but impose significant computational and communication overhead, making them unsuitable for resource-constrained settings. Alternative solutions using differential privacy [30] sacrifice accuracy.

In this work, we propose FedMod, a privacy-preserving VFL framework based on additive n-out-of-n secret sharing. Unlike cryptographic or differential privacy techniques, FedMod achieves privacy without compromising performance or accuracy. Moreover, FedMod's lightweight computation makes it well-suited for devices with limited processing power, such as IoT devices. In particular, our contributions are:

- We introduce FedMod, a vertical federated learning framework that achieves privacy through lightweight n-out-of-n additive secret sharing, supported by a scalable multi-server architecture that eliminates the need for encryption, trusted execution environments, or key management.
- We design secure and efficient protocols for both regression and classification tasks in VFL settings.
- We evaluate FedMod experimentally and show it achieves accuracy comparable to centralized models, with significantly reduced computation and communication costs.

2 Background

This section presents some preliminaries to understand the proposed framework.

2.1 Federated Learning

FL enables collaborative model training without sharing raw data. It is typically categorized into Horizontal Federated Learning (HFL) and Vertical Federated Learning (VFL) [35].

HFL applies when datasets across parties share similar features but differ in user identities. Two widely used HFL approaches are Federated Stochastic Gradient Descent (FedSGD) and Federated Averaging (FedAvg) [18]. In FedSGD, parties compute gradients on their local data and send them to a central server for aggregation. In FedAvg, parties train local models and send updated parameters to the server for averaging.

VFL, on the other hand, is suitable when parties share user identities but hold different features. For instance, banks and regulatory agencies may have overlapping customers but different financial information. By jointly training a model using VFL [32], they can detect activities like money laundering without exposing sensitive data. VFL is particularly relevant for applications requiring cross-organization collaboration on vertically partitioned datasets.

2.2 Cryptography

Cryptographic techniques such as Homomorphic Encryption (HE) and Functional Encryption (FE) have been applied in VFL to protect data during computation. HE enables arithmetic operations on ciphertexts, allowing model training on encrypted data [8,10], while FE permits evaluating specific functions on encrypted data without revealing the input [3,32].

Despite their strong privacy guarantees, these techniques introduce significant computational and communication overhead [23]. Complex operations in HE often require approximations, potentially reducing model accuracy and increasing training time.

2.3 Secret Sharing

Secret sharing divides a value into multiple random shares such that the original value can only be reconstructed when all shares are combined [23]. Among various schemes, additive secret sharing is one of the most lightweight and practical approaches [31], particularly in federated learning scenarios where efficiency is crucial.

In this work, we adopt a standard n-out-of-n additive secret sharing [16] scheme. Each intermediate output is split into n shares, with the first $n-1$ chosen at random and the last computed to preserve the sum. This approach is computationally efficient and provides strong privacy, as no subset of fewer than n shares reveals any information about the original value. Unlike encryption-based techniques, it avoids the need for complex key management or expensive ciphertext operations.

3 Related Work

This section outlines related work in privacy-preserving federated learning, focusing on perturbation, cryptographic, and secret sharing methods.

3.1 Perturbation

Differential Privacy (DP) is widely used to protect sensitive data by adding noise to intermediate computations. While effective in limiting information leakage, DP often reduces model accuracy and requires delicate tuning of the noise scale. Several VFL approaches such as VAFL [4], HDVFL [27], and PPAVFL [9] apply perturbation for privacy protection. CZOFO [29] improves over traditional DP by integrating zeroth- and first-order optimization techniques, but remains sensitive to hyperparameter selection and can incur additional communication cost.

3.2 Cryptographic Primitives

Cryptographic techniques like Homomorphic Encryption (HE) and Functional Encryption (FE) are also used to secure federated computations. PrivFL [17], Hardy [11], Quasi [33], FedV [32], MVFLS [23], EFMVFL [13], and FedVS [15] are notable examples. While HE enables arithmetic over encrypted data and FE provides fine-grained control, these methods introduce substantial computational overhead and are ill-suited for resource-constrained environments [6]. Recent work like RLSA-PFL [25] proposes a lightweight cryptographic protocol for secure aggregation and model inconsistency detection. However, it targets horizontal federated learning and still depends on cryptographic setups and MAC verification, making it less practical for vertical collaboration where low-overhead solutions are preferred.

3.3 Secret Sharing

Secret sharing avoids the computational cost of encryption while ensuring strong privacy guarantees. It divides sensitive values into random shares distributed across multiple entities. Approaches like SecAGG [2], PPAVFL [9], MVFLS [23], EFMVFL [13], and FedVS [15] use single-server or no-server variants, with some also incorporating cryptographic layers [19]. These models face scalability and performance challenges, particularly in VFL with growing party numbers. Fed-Share [7] addresses this by using an n-out-of-n additive secret-sharing strategy with multiple aggregation servers. Building on this concept, our proposed FedMod framework removes encryption entirely and applies multi-server additive secret sharing in a vertical setting, offering a scalable and efficient solution that simplifies deployment, reduces communication bottlenecks, and retains model accuracy.

4 System Overview

Our framework consists of three main components as shown in 1: (1) m parties, each owning a disjoint set of features for a common set of z aligned samples. Parties compute intermediate outputs locally, split them into n additive shares, and distribute them to n middle servers. (2) Each middle server receives one share per party, aggregates them, and forwards the result to the leading server, which orchestrates training by combining outputs from all middle servers. The leading server never sees raw data or individual contributions. Real-world analogs include healthcare institutions acting as aggregators and ministers of health serving as middle servers.

We adopt a semi-honest threat model: entities follow the protocol but may try to infer private information. Assumptions include: (i) the leading server does not collude [14,34]; (ii) at least one middle server is honest; (iii) entity alignment is completed beforehand. Three adversarial scenarios are considered: (1) *Semi-honest behavior*: middle servers see only random shares, revealing nothing individually. (2) *Partial collusion among middle servers*: up to $n-1$ colluding servers still cannot reconstruct intermediate outputs. (3) *Collusion between middle servers and parties*: even $m-1$ colluding parties and $n-1$ servers cannot infer another party's data without full reconstruction. Shares are generated and reconstructed over a finite field $\mathbb{Z}_K$, ensuring that no subset of parties or servers can infer private values without all n shares. Privacy holds unless full collusion occurs, which is outside the scope of this work.

5 FedMod

FedMod is a privacy-preserving approach for vertical federated learning, enabling model $\mathcal{M}$ training on dataset $\mathcal{D}$ from party set P without data leakage. In regression, the first party (p_1) holds labels, while in classification, the leading server does. Training consists of forward and backward steps, repeated for each record until completion.

5.1 Forward

The whole feature set is $X = \{x_1, x_2, ..., x_d\}$ where d is the number of all features. There are m numbers of parties. Each party (p_i) has access to a subset of features and weights following Eq. 1 and Eq. 2, respectively.

$$X_i = \begin{cases} x_{(i-1)\times \lfloor \frac{d}{m} \rfloor +1}, ..., x_{(i-1)\times \lfloor \frac{d}{m} \rfloor + \lfloor \frac{d}{m} \rfloor} & \text{if } 1 \leq i < m \\ x_{(i-1)\times \lfloor \frac{d}{m} \rfloor +1}, ..., x_{(i-1)\times \lfloor \frac{d}{m} \rfloor + \lfloor \frac{d}{m} \rfloor + (d\%m)} & \text{if } i = m \end{cases} \quad (1)$$

$$W_i = \begin{cases} w_{(i-1)\times \lfloor \frac{d}{m} \rfloor +1}, ..., w_{(i-1)\times \lfloor \frac{d}{m} \rfloor + \lfloor \frac{d}{m} \rfloor} & \text{if } 1 \leq i < m \\ w_{(i-1)\times \lfloor \frac{d}{m} \rfloor +1}, ..., w_{(i-1)\times \lfloor \frac{d}{m} \rfloor + \lfloor \frac{d}{m} \rfloor + (d\%m)} & \text{if } i = m \end{cases} \quad (2)$$

where, i is the index of parties, m is the total number of parties, d is the number of all features, and % represents modulo operation. There is no common feature between parties ($\bigcap_{i=1}^{i=m} X_i = \emptyset$), and the union of all features makes up the total feature set($\bigcup_{i=1}^{i=m} X_i = X$). Each party (p_i) does a forward pass on its feature set (X_i) using the initial weights set (W_i) and a bias (b_i) so that it generates its intermediate output. The bias is different for each party. Therefore, we have a set of biases for all parties ($B = \{b_1, b_2, ..., b_m\}$). In addition, we have a set of labels for all z samples ($Y = \{y_1, y_2, ..., y_z\}$). The whole process for the forward pass is shown in Algorithm 1 and demonstrated in Fig. 1.

Each party will calculate its intermediate output ($\tilde{X}_i$) for each record, which is the sum of its weights multiplied by the respective feature values plus the bias. The intermediate output of the first party is different than the intermediate outputs of the other parties since the first party holds the labels. After the intermediate outputs are created, then each party i needs to create n shares such that each share will be shared with the corresponding middle server.

Each party uses n-out-of-n additive secret sharing to split $\tilde{X}_i$ into n shares. The first $n-1$ shares are sampled uniformly at random, and the final share is

Algorithm 1. Forward

1: **Input:** Data matrix for party i (X_i), Weights matrix for party i (W_i), bias for party i (b_i), index of a record (r), number of parties (m), number of middle servers (n), set of labels of all records (Y), address of middle server i (s_i), Address of leading server (S^*), additive shares of party i for middle server j (sh_i^j), secret key shared between parties and leading server (K)
2: **Parties:**
3: **for** each party $i = 1, 2, \ldots, m$ **do**
4:    **if** $i = 1$ **then**
5:       $\tilde{X}_i \leftarrow \vec{X}_i^r \cdot \vec{W}_i^T + b_i - Y_r$
6:    **else**
7:       $\tilde{X}_i \leftarrow \vec{X}_i^r \cdot \vec{W}_i^T + b_i$
8:    **end if**
9:    Generate $n - 1$ random shares $sh_i^1, \ldots, sh_i^{n-1}$
10:   $sh_i^n \leftarrow \tilde{X}_i - \sum_{j=1}^{n-1} sh_i^j \mod K$
11:   **for** $l = 1, 2, \ldots, n$ **do**
12:      send sh_i^l to s_l
13:   **end for**
14: **end for**
15: **Middle servers:**
16: **for** each middle server $l = 1, 2, \ldots, n$ **do**
17:   $sum_l \leftarrow \sum_{i=1}^{m} sh_i^l$
18:   send sum_l to S^*
19: **end for**
20: **Leading server:**
21: $total_r \leftarrow \sum_{l=1}^{n} sum_l$
22: $e \leftarrow total_r \mod K$

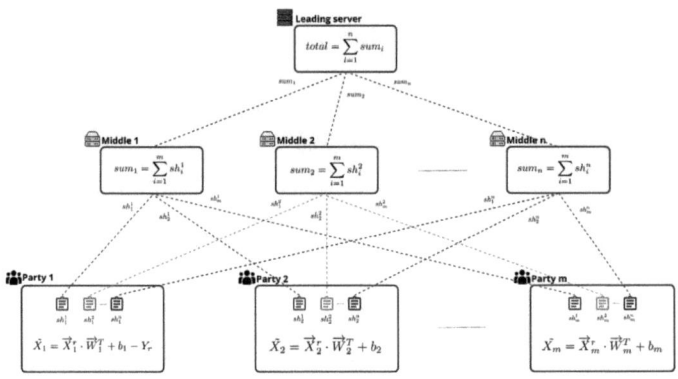

Fig. 1. Forward

computed so that their sum equals $\tilde{X}_i$. Each share is then sent to a corresponding middle server s_j. This ensures that no single server can learn anything about the intermediate output unless all n shares are combined.

According to Algorithm 1, parties send secret shares to all middle servers, ensuring each middle server s_l receives one share per party. These shares reveal no information since individual shares are meaningless. The middle servers only sum the received values, preventing the leading server from distinguishing individual contributions. Consequently, the leading server can derive only the aggregated intermediate outputs, preserving party-level privacy. Upon receiving data from all middle servers, the leading server sums them and obtains the total intermediate output of all parties. Each middle server i sends the sum of its m shares, and the leading server aggregates the values from all n middle servers as $total = \sum_{i=1}^{n}(sh_i^1 + \cdots + sh_i^m)$.

For example, if the leading server applies the modulo operation on data sample r of the dataset $\mathcal{D}$, it will get $total_r$ in Eq. (3) for the data sample r.

$$total_r = \sum_{i=1}^{m}((X_i^r \cdot W_i^T) + b_i) - Y_r \tag{3}$$

When $total_r$ is calculated, we can use it to compute the Mean Squared Error (MSE), defined as MSE $= (total_r)^2$. In linear regression, this loss is used to update the weights and biases. Parties need to calculate the derivative of the MSE with respect to the weight and bias according to Eq. 4.

$$\frac{\partial \text{MSE}}{\partial W} = total_r \cdot X_r, \quad \frac{\partial \text{MSE}}{\partial b} = total_r \tag{4}$$

Based on Eq. (4), the parties need the $total_r$ value to update their weights and biases. We will call the $total_r$ value the error. The leading server will send this error value back to the parties in a private way based on the approach that will be discussed in the following section.

5.2 Backward

In the second step of the training process, the leading server sends back the error to the parties in a private way using the middle servers, so that the parties can update their weights and biases accordingly. Algorithm 2 shows the whole process for backward propagation.

The leading server creates n additive shares of the error value so that no entity other than the parties can know it. Since the error is the same for all parties, each party reconstructs it by summing all the shares: $e = \sum_{i=1}^{n} e_i \mod K$.

Each share e_i is sent to a different middle server, which then forwards it to all parties. Each party aggregates the received n shares to reconstruct the error value. Once the error is recovered, each party computes the gradients and applies weight and bias updates locally. W^* and b^* are the updated weights and biases, λ is the regularization parameter, and $R(.)$ is the regularization term. This iterative training process requires collaboration from all parties in every round.

5.3 Classification

We handle binary classification with a sigmoid function and multi-class classification using the One-vs-All (OvA) approach [21], assuming the leading server holds the labels.

Algorithm 2. Backward

1: **Input:** error value (e), data matrix for party i (X_i), index of a record (r), address of middle server i (s_i), regularization item ($R(.)$), regularization parameter (λ), learning rate (η), secret key shared between parties and leading server (K)
2: **Output:** update weights (W) and bias (b) in parties based on the error value
3: **Procedure** Backward()
4: **Leading server:**
5: Generate $n-1$ random shares $e_1, ..., e_{n-1}$
6: $e_n \leftarrow e - \sum_{j=1}^{n-1} e_j \mod K$
7: **for** $i = 1, 2, ..., n$ **do**
8:     Send e_i to s_i
9: **end for**
10: **Middle servers:**
11: **for** each server $i = 1, 2, ..., n$ **do**
12:     Send e_i to all parties
13: **end for**
14: **Parties:**
15: **for** each party $j = 1, 2, ..., m$ **do**
16:     $e \leftarrow \sum_{i=1}^{n} e_i \mod K$
17:     $\frac{\partial loss}{\partial W_j} \leftarrow e \times X_j^r$
18:     $\frac{\partial loss}{\partial b_j} \leftarrow e$
19:     $W_j^* \leftarrow W_j - \eta \times (\frac{\partial loss}{\partial W_j} + \lambda R(W_j))$
20:     $b_j^* \leftarrow b_j - \eta \times (\frac{\partial loss}{\partial b_j} + \lambda R(b_j))$
21: **end for**

Binary Classification: Similar to regression, each party computes $(X_i^r W_i^T + b_i)$, and the leading server aggregates to compute $total_r = \sum_{i=1}^{m}((X_i^r W_i^T) + b_i)$. Unlike regression, Y_r is not subtracted immediately. Instead, the leading server applies the sigmoid function $\hat{X} = \frac{1}{1+e^{-total_r}}$, interprets it as the predicted probability, and computes the error $\hat{X} - Y_r$, which is sufficient for gradient updates. Although Binary Cross Entropy (BCE) loss is standard:

$$\text{Loss}_{\text{BCE}} = -\left[Y_r \log(\hat{X}) + (1 - Y_r)\log(1 - \hat{X})\right],$$

its derivative simplifies to:

$$\frac{\partial \text{Loss}_{\text{BCE}}}{\partial W} = (\hat{X} - Y_r)X_r, \quad \frac{\partial \text{Loss}_{\text{BCE}}}{\partial b} = (\hat{X} - Y_r),$$

matching the regression update rule (Eq. 4), so the backward flow remains unchanged.

Multi-class Classification: For q classes, the OvA scheme trains q parallel sigmoid classifiers. Each party holds q weight-bias pairs, producing q outputs sent to the leading server. The server applies sigmoid on each, chooses the max for inference, and returns q errors to update party parameters. This design enables efficient multi-class training while preserving FedMod's modular forward and backward logic.

6 Evaluation

In this section, we will detail the experimental setup and the results. Code

6.1 Experimental Setup

Experiments were run on a macOS 14.4.1 system (2.6 GHz 6-core Intel i7, 16GB RAM) using Python 3.9.19 and TensorFlow 2.16.1.

Datasets. Our study leverages a variety of datasets to evaluate the performance and robustness of the proposed models. Table 1 provides a summary of these datasets, including their size, number of features, and type.

Table 1. Summary of datasets used in evaluation.

Dataset	# Samples	# Features	Task Type
Heart Disease [26]	1025	14	Binary Classification
Ionosphere [24]	351	34	Binary Classification
Parkinson [22]	756	754	Binary Classification
Phishing [1]	1353	9	Multi-Class Classification

Benchmarking. We compare FedMod with a baseline VFL model (no security), and state-of-the-art approaches including homomorphic encryption (HE), differential privacy (DP), FedV [32], and CZOFO [29]. All models use the same architecture (sigmoid activation, no hidden layers) to ensure consistent and fair comparisons focused on speed and communication cost. Since FedV lacks a public implementation, we reproduced it per its description, including its functional encryption (FE) component using the Cryptography library [5]. We used elliptic curve keys (SECP256R1) and derived shared keys. FedMod outperformed FedV in both speed and accuracy. DP was implemented using the Laplace mechanism ($\epsilon = 0.5$, sensitivity $= 1$), which reduced accuracy due to noise but improved runtime over HE. HE was implemented using TenSEAL with polynomial modulus degree 2048, coefficient sizes [20, 20], global scale 2^{10}, and Galois keys. While HE achieved good accuracy, it incurred prohibitive computational costs. For CZOFO, we used its official code [28]. While it excels at minimizing communication cost, FedMod achieved consistently lower training time, highlighting its efficiency and lightweight design. Lack of available implementations limited further comparisons.

6.2 Experimental Results

Accuracy and Time Comparison. One of the primary contributions of this paper is the demonstration that our approach, FedMod, can achieve accuracy

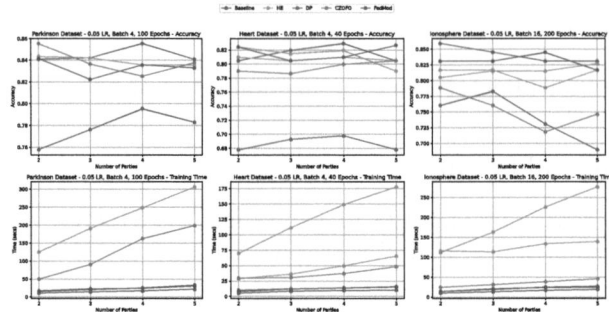

Fig. 2. Accuracy and training time comparison across different datasets and methods: FedMod, Baseline, DP, FedV, HE, and CZOFO.

comparable to the baseline VFL setting, which operates without security measures, while significantly reducing training time compared to other security-focused approaches.

Figure 2 highlights FedMod's balance between accuracy and training time across datasets. FedMod achieves near-baseline accuracy while significantly reducing training time compared to encryption-based methods like FedV and HE, which suffer from computational overhead. While DP matches FedMod in speed, it sacrifices accuracy due to noise in intermediate outputs. HE and FedV have the longest training time, with FedV excluded from the Parkinson dataset due to excessive computation. CZOFO balances accuracy and training time but remains slower than FedMod, especially in smaller batches. These results underscore FedMod's efficiency, avoiding encryption overhead and outperforming architectures like CZOFO, which rely on Zero-Order Optimization (ZOO).

Communication Cost. Communication cost is a key factor in federated learning, especially in real-world scenarios requiring frequent interactions. Minimizing data transfer is crucial for efficiency and scalability. This section compares the communication cost of FedMod with CZOFO, a recent method that enhances efficiency through advanced compression techniques.

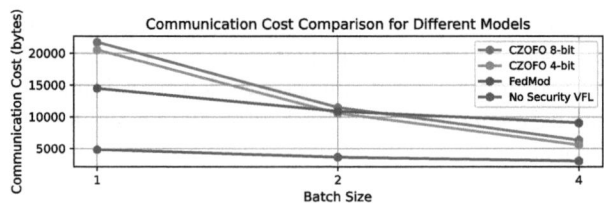

Fig. 3. Comparison of communication costs in bytes for one epoch of training on the Parkinson dataset between FedMod, CZOFO (with 4-bit and 8-bit compression), and a baseline VFL model without security.

We assess communication cost by measuring data transfer per epoch on the Parkinson dataset, ensuring a standardized comparison. The x-axis in Fig. 3 represents batch size, a crucial factor since CZOFO excels with larger batches. We compare FedMod with CZOFO due to its strong communication efficiency among similar methods. Since FedMod avoids costly encryption techniques like HE or FE, its communication cost remains inherently lower.

Figure 3 presents the communication cost for FedMod, the baseline model without security, and CZOFO with 4-bit and 8-bit compression. These settings balance compression efficiency and accuracy, as smaller bit sizes reduce accuracy, while larger sizes offer diminishing benefits. FedMod achieves lower communication cost than CZOFO with smaller batches, despite its additional middle servers. This highlights FedMod's efficiency in minimizing overhead while maintaining competitive communication cost. Reported costs include all bandwidth usage among entities.

Ablation Study. A key strength of FedMod is its scalability. Unlike methods restricted to two-party settings, FedMod supports multiple parties and additional middle servers for enhanced security. This flexibility makes it suitable for real-world scenarios with varying participants. We compare FedMod to a baseline VFL without security, showing it maintains similar accuracy and competitive timing as the number of parties increases.

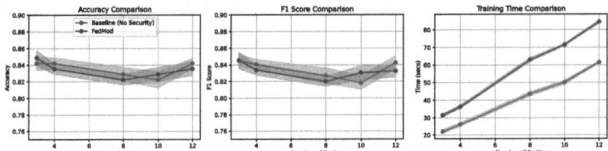

Fig. 4. Accuracy, F1 Score, and Time comparison with selected number of parties on the Parkinson dataset. FedMod results are based on a 2 middle servers setting. Shaded regions indicate the standard deviation over multiple runs.

Figure 4 shows that FedMod maintains strong accuracy, F1 score, and timing across different numbers of parties. While computation time increases slightly due to testing on a single system, in real-world distributed settings, it is expected to remain close to the two-party case. This implementation setup aligns with other related works that have also conducted their experiments on a single device [29]. This demonstrates FedMod's scalability, ensuring robust performance and security even with many parties.

Additionally, we examine the impact of varying the number of middle servers on the performance of FedMod. Since altering the number of middle servers does not affect the results of the No Security VFL scenario, we compare FedMod's performance in a two-middle server setting with No Security VFL in a two-party setting.

Table 2. Performance comparison as number of middle servers increases (2-party setting, Parkinson dataset).

Method	Accuracy	F1 Score	Time (s)
Baseline (No Security)	0.8421	0.8456	10.77
FedMod (2 servers)	0.8421	0.8456	17.37
FedMod (3 servers)	0.8158	0.8129	18.11
FedMod (4 servers)	0.8355	0.8360	19.98
FedMod (8 servers)	0.8553	0.8619	22.65
FedMod (10 servers)	0.8421	0.8456	25.02
FedMod (12 servers)	0.8355	0.8360	29.32

Table 2 shows that the number of middle servers has minimal impact on the model's performance. There is a slight increase in training time as the number of middle servers rises. This is expected given that the experiments were conducted on a single device. In real-world applications, where tasks are distributed across multiple devices, this increase in time is likely to be negligible. This reinforces the scalability of FedMod, demonstrating that it can accommodate additional security layers without compromising performance.

We also evaluated the stability and convergence of learning curves across methods. As shown in Fig. 5, FedMod closely tracks No Security VFL in both stability and convergence. CZOFO is more sensitive to learning rate, HE shows fluctuations due to encrypted precision loss, and DP exhibits the highest instability from added noise. FedMod achieves reliable convergence, making it a strong choice for secure VFL.

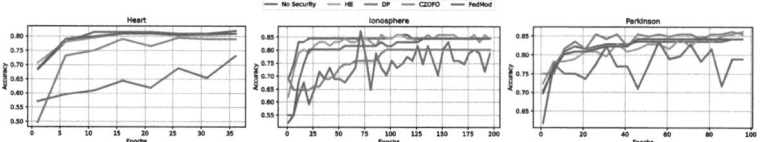

Fig. 5. Learning curves comparison across different datasets and methods: FedMod, No Security, DP, HE, and CZOFO. Lines are plotted using values from every 5 epochs.

7 Security Discussion

Our framework uses n-out-of-n additive secret sharing, which ensures that no information is revealed unless all shares are combined. All operations involving shares are performed over a finite field $\mathbb{Z}_K$, where K is a sufficiently large prime number chosen to ensure correctness, uniform randomness, and security under modular arithmetic. This guarantees that share generation and reconstruction behave correctly and securely.

We analyze its resilience under three threat models:

(1) *Semi-Honest Model:* All parties follow the protocol but may attempt passive inference. Each middle server sees only one random, meaningless share, and the leading server receives only aggregated results.
(2) *Collusion Among Middle Servers:* Up to $n-1$ colluding middle servers cannot reconstruct the original value without the final share. This guarantee is information-theoretic and holds regardless of computational power.
(3) *Collusion Between Middle Servers and Parties:* Collusion among fewer than n middle servers and some parties still fails to breach privacy, as reconstruction requires all shares. Thus, no subset of colluding entities can infer private data without full collusion.

Our approach avoids encryption and key management, relying on the strength of additive secret sharing. Privacy can be tuned by increasing the number of middle servers, staying secure as long as one remains honest.

8 Conclusion

This paper introduced FedMod, a lightweight and scalable framework for secure vertical federated learning based on multi-server additive secret sharing. FedMod ensures strong privacy guarantees without relying on encryption or heavy computation, offering a compelling trade-off between security, efficiency, and accuracy. Extensive experiments demonstrate that FedMod maintains performance close to unsecured baselines while significantly reducing overhead compared to existing secure VFL methods. Future work will explore integrating FedMod with more complex models, such as deep neural networks, to extend its applicability to large-scale and high-dimensional real-world tasks.

Acknowledgments. This research is supported by the Natural Sciences and Engineering Research Council of Canada (NSERC) Discovery Grant (RGPIN-2019-05689).

References

1. Abdelhamid, N.: Website Phishing. UCI Machine Learning Repository (2016)
2. Bonawitz, K., et al.: Practical secure aggregation for privacy-preserving machine learning. In: Proceedings of the 2017 ACM SIGSAC Conference on Computer and Communications Security, pp. 1175–1191 (2017)
3. Boneh, D., Sahai, A., Waters, B.: Functional encryption: definitions and challenges. In: Ishai, Y. (ed.) TCC 2011. LNCS, vol. 6597, pp. 253–273. Springer, Heidelberg (2011). https://doi.org/10.1007/978-3-642-19571-6_16
4. Chen, T., Jin, X., Sun, Y., Yin, W.: VAFL: a method of vertical asynchronous federated learning. arXiv preprint arXiv:2007.06081 (2020)

5. cryptography developers, T.: cryptography (2024). https://pypi.org/project/cryptography/, version 41.0.3
6. Dhanda, S.S., Singh, B., Jindal, P.: Lightweight cryptography: a solution to secure IoT. Wireless Pers. Commun. **112**(3), 1947–1980 (2020)
7. Fazli Khojir, H., Alhadidi, D., Rouhani, S., Mohammed, N.: FedShare: secure aggregation based on additive secret sharing in federated learning. In: Proceedings of the 27th International Database Engineered Applications Symposium, pp. 25–33 (2023)
8. Gong, M., et al.: A multi-modal vertical federated learning framework based on homomorphic encryption. IEEE Transactions on Information Forensics and Security (2023)
9. Gu, B., Xu, A., Huo, Z., Deng, C., Huang, H.: Privacy-preserving asynchronous vertical federated learning algorithms for multiparty collaborative learning. IEEE Trans. Neural Netw. Learn. Syst. **33**(11), 6103–6115 (2021)
10. Guo, Y., et al.: Efficient and privacy-preserving federated learning based on full homomorphic encryption. arXiv preprint arXiv:2403.11519 (2024)
11. Hardy, S., et al.: Private federated learning on vertically partitioned data via entity resolution and additively homomorphic encryption. arXiv preprint arXiv:1711.10677 (2017)
12. of Health, U.D., Services, H.: HIPAA privacy rule. https://www.hhs.gov/hipaa/for-professionals/privacy/index.html (2024). Accessed 18 June 2024
13. Huang, Y., et al.: EFMVFL: an efficient and flexible multi-party vertical federated learning without a third party. ACM Trans. Knowl. Discov. Data **18**(3), 1–20 (2023)
14. Li, Q., et al.: Vertical federated learning: taxonomies, threats, and prospects. arXiv preprint arXiv:2302.01550 (2023)
15. Li, S., Yao, D., Liu, J.: FedVS: straggler-resilient and privacy-preserving vertical federated learning for split models. In: International Conference on Machine Learning, pp. 20296–20311. PMLR (2023)
16. Malkin, T.: Lecture notes on secret sharing. https://www.cs.columbia.edu/~tal/4261/F19/secretsharingf19.pdf (2019), cSOR 4231: Computer Security, Columbia University, Fall 2019
17. Mandal, K., Gong, G.: PrivFL: practical privacy-preserving federated regressions on high-dimensional data over mobile networks. In: Proceedings of the 2019 ACM SIGSAC Conference on Cloud Computing Security Workshop, pp. 57–68 (2019)
18. McMahan, B., Moore, E., Ramage, D., Hampson, S., y Arcas, B.A.: Communication-efficient learning of deep networks from decentralized data. In: Artificial Intelligence and Statistics, pp. 1273–1282. PMLR (2017)
19. Natarajan, D., Dai, W.: Seal-embedded: a homomorphic encryption library for the internet of things. In: IACR Transactions on Cryptographic Hardware and Embedded Systems, pp. 756–779 (2021)
20. Office of the Privacy Commissioner of Canada: Personal Information Protection and Electronic Documents Act (PIPEDA) (2024)
21. Rifkin, R., Klautau, A.: In defense of one-vs-all classification. J. Mach. Learn. Res. **5**, 101–141 (2004)
22. Sakar, C.O., et al.: A comparative analysis of speech signal processing algorithms for Parkinson's disease classification and the use of the tunable q-factor wavelet transform. Appl. Soft Comput. **74**, 255–263 (2019)
23. Shi, H., Jiang, Y., Yu, H., Xu, Y., Cui, L.: MVFLS: multi-participant vertical federated learning based on secret sharing. The Federate Learning, pp. 1–9 (2022)

24. Sigillito, V., Wing, S., Hutton, L., Baker, K.: Ionosphere. UCI Machine Learning Repository (1989). https://doi.org/10.24432/C5W01B
25. Sultan, N.H., et al.: RLSA-PFL: Robust lightweight secure aggregation with model inconsistency detection in privacy-preserving federated learning. arXiv preprint arXiv:2502.08989 (2024)
26. Unknown: Heart disease dataset. https://www.kaggle.com/datasets/johnsmith88/heart-disease-dataset (2024). Accessed 30 May 2024
27. Wang, C., Liang, J., Huang, M., Bai, B., Bai, K., Li, H.: Hybrid differentially private federated learning on vertically partitioned data. arXiv preprint arXiv:2009.02763 (2020)
28. Wang, G.: CZOFO GitHub page. https://github.com/GanyuWang/VFL-CZOFO (2023). Accessed 24 Aug 2024
29. Wang, G., Gu, B., Zhang, Q., Li, X., Wang, B., Ling, C.X.: A unified solution for privacy and communication efficiency in vertical federated learning. Adv. Neural Inf. Process. Syst. **36** (2024)
30. Wei, K., et al.: Federated learning with differential privacy: algorithms and performance analysis. IEEE Trans. Inf. Forensics Secur. **15**, 3454–3469 (2020)
31. Xiong, L., Zhou, W., Xia, Z., Gu, Q., Weng, J.: Efficient privacy-preserving computation based on additive secret sharing. arXiv preprint arXiv:2009.05356 (2020)
32. Xu, R., Baracaldo, N., Zhou, Y., Anwar, A., Joshi, J., Ludwig, H.: FedV: privacy-preserving federated learning over vertically partitioned data. In: Proceedings of the 14th ACM Workshop on Artificial Intelligence and Security, pp. 181–192 (2021)
33. Yang, K., Fan, T., Chen, T., Shi, Y., Yang, Q.: A quasi-newton method based vertical federated learning framework for logistic regression. arXiv preprint arXiv:1912.00513 (2019)
34. Ye, M., Shen, W., Snezhko, E., Kovalev, V., Yuen, P.C., Du, B.: Vertical federated learning for effectiveness, security, applicability: A survey. arXiv preprint arXiv:2405.17495 (2024)
35. Zhang, C., Xie, Y., Bai, H., Yu, B., Li, W., Gao, Y.: A survey on federated learning. Knowl.-Based Syst. **216**, 106775 (2021)

Throughput-Driven Database Replication Using a Ring-Based Order Protocol

Ye Liu[1]( ), Paul Ezhilchelvan[1], Yingming Wang[1], and Jim Webber[2]

[1] School of Computing, Newcastle University, Newcastle upon Tyne NE4 5TG, UK
{Y.Liu197,Paul.Ezhilchelvan,Y.Wang303}@Newcastle.ac.uk
[2] Neo4j UK, London SE1 0LH, UK
Jim.Webber@Neo4j.com

Abstract. We present a database replication architecture that guarantees ACID transaction properties as well as high throughput expected of modern database systems. Higher throughput results due to server replicas processing distinct, non-overlapping subsets of incoming transactions *in parallel*. Our novel approach addresses all challenges that emerge in ensuring ACID properties across *all* incoming transactions processed in parallel even when access pattern of transactions is not known *a priori*. At the core of our approach is a high-throughput, ring-based total order protocol which the database replicas use to reach consensus for resolving conflicts among transactions, ensuring serializability and accomplishing atomic commit. After presenting the architecture, protocol performance is evaluated through implementations when replication degree is two and three, tolerating at most one replica crash. While 2-fold replication requires perfect crash detection, three-fold can do with weak detectors.

1 Introduction

Replicating a database for high availability has long been studied, implemented, and analysed for various performance characteristics (see [13]). Replication that also ensures ACID properties needs to address greater, additional challenges. *Atomicity (A), Consistency (C), Isolation (I), Durability (D)* are the ACID properties that ensure total integrity at the point of transaction termination, despite host crashes and transactions seeking to access common data items in an order incompatible with C and I properties. For example, the problem of 'incompatible access' or *conflict* becomes more challenging to solve when database replicas process distinct transactions in parallel. Consequently, several performance studies, e.g. [18], shows that ACID replications offer a much smaller throughput even at medium loads, compared to non-ACID ones; however, the latter permit database replica states to diverge and hence require state reconciliation which can be next to impossible in some database contexts [6].

This paper addresses the challenge of improving throughput for ACID replication systems. Our approach involves fully replicating a database on multiple servers that initially execute distinct subsets of input transactions in parallel, but finally generate identical transaction outcomes and commit identical state updates. At the heart of our proposal is a high-throughput ring-based total order protocol supporting three essentials: event ordering required for server replication, identical inter-replica concurrency control to ensure C and I properties of ACID, and *Two-Phase Commit* (2PC) [9] to guarantee A and D.

While our approach is novel when transaction access pattern is initially unknown, it derives its theoretical underpinning from three canonical findings: (i) total ordering of transactions (or *Atomic Broadcast*) and solving *Consensus* (to resolve conflicting transactions) are reducible to each other under crash failures [2] (ii) the two-phase commit (2PC) is only a simplified instance of consensus [10], and (iii) maximum throughput is achieved when transaction ordering is done over a logical, unidirectional ring network [11].

Consider, as a motivating example, a database replicated on n fail-independent servers: $\{R_i, R_j | 1 \leq i, j \leq n, n \geq 2\}$. Say, concurrent transactions T_i^k and $T_i^{k'}$ execute in R_i and access overlapping sets of data items. This will be termed as a *local* conflict. To ensure C and I, R_i needs to resolve who waits or gives up for whom, and this can be done autonomously within R_i. Let us say transaction T_j^l executes in R_j in parallel and wishes to access replicas of data items common with T_i^k executing simultaneously in R_i. This is termed as a *global* conflict which can be detected and be resolved only after parallel executions in R_i and R_j are over. A total order (or simply an order) protocol takes inputs from distributed servers and lets all servers decide identically on an order over the combined set of inputs. It thus provides precedence on transactions for replicas to identically decide which of the globally conflicting transactions waits or gives up.

The paper is organised as follows. After presenting related work below, Sect. 2 presents our approach together with necessary background. It also presents another popular approach to highlight the novelty of ours. Section 3 describes the role of our order protocol in building a single server abstraction wherein all replicas process all transactions identically (commit or abort) even though they start off processing distinct streams of incoming transactions. Section 4 implements the ring-based protocol and measures its responsiveness. This protocol has been presented earlier in [14] together approximations for response time estimation. The measured results are shown to be close to the estimates, allowing us to dynamically adapt protocol parameters for real world situations. Finally, concluding remarks are provided in Sect. 5.

1.1 Related Work

A recent book [13] compiles the vast material addressing database replication. [12] introduces a BDI-based framework that improves replication by adaptive behaviour and formal verification. The paper [17] classifies replication techniques using three parameters, namely: active vs passive replication, replica interactions per transaction vs per operation, and non-voting vs voting based decision on transaction commit or abort. We use parallelised passive replication with each replica acting as the primary for distinct input streams, per transaction interactions, and non-voting for commit/abort decisions.

The performance study in [18] (referred to earlier) compares five ordering-based replication techniques; lazy replication, where consensus is not applied to resolve conflicts, offers high throughput at a high risk of state divergence for Graph databases as argued in [5]. The order protocol used in [18] for ACID replication is leader-based where the leader is a performance bottleneck. Our ring-based protocol is leader-free - an important feature that allows it to achieve the highest possible throughput [11].

2 Our Approach

2.1 Conflict and Concurrency Control

Concurrent transactions accessing common data items in a database system can create *conflicts*. There are three types of access conflict: after an ongoing transaction, say T_i, has written a data item, say, X, if another one, say T_j, wishes to write or read X, then a *write-write* or *write-read* conflict is said to occur, respectively; similarly, a *read-write* conflict arises when T_j seeks to write X after T_i has read X. Suppose also that T_j has already accessed another data item Y and T_i (after having accessed X) seeks a conflicting access on Y; an incompatibility with C and I guarantees would arise if both T_i and T_j were allowed to go ahead ignoring these access conflicts completely: T_i would have accessed X before T_j with T_j accessing Y before T_i. This would violate C and I which require transactions access *all* common data items in the same order.

The literature proposes many *isolation levels* [15], from the most restricted to unrestricted, to guarantee C and I. The unrestricted *serializability* must eliminate all incompatibility that can arise due to any of three types of conflicts discussed earlier. It will be our target isolation level here.

Irrespective of the isolation level sought, there are broadly two ways to resolve a conflict: *Wait* and *Abort*. In the former, later transaction(s) wait until the earlier transaction completes its execution. Assuming that T_i and T_j are executing in the same replica, T_j will wait to access X until T_i completes while it would be the other way around on Y. So, the wait strategy must be accompanied by deadlock detection and avoidance strategies, e.g., the work from [4] maintain an access graph for ensuring the smallest possible abort ratio.

In *Abort* approach, one transaction (T_i or T_j) aborts itself after a limited or no waiting. We use here *No-Wait, Instantaneous Abort* wherein a transaction encountering a conflict will instantly abort itself. Thus, if both T_i and T_j are running on the same replica, their access conflicts would be *local* and can lead to both aborting as they individually encounter a conflict. Contrary to the intuition that *Instantaneous Abort* may cause too many transactions to abort, our earlier evaluations (see [7]), both model- and implementation-based, demonstrate that the abort ratio is acceptably small and not considerably larger than that the smallest returned by [4]. *Instantaneous Abort*, on the other hand, eliminates deadlock possibilities, extracts near-zero implementation overhead in replicated systems and also is the most throughput-friendly.

If T_i and T_j are running on different replicas, say in R_i and R_j respectively, their conflicts would be *global* and not detectable during their parallel execution. Post-execution, if the order protocol orders, say, T_i before T_j then all n replicas will deem T_j to be aborted and accept the outcomes of T_i execution in R_i.

2.2 Architectures of Replicated Database System

We present our architecture after presenting one of the most popular replication architectures found in the literature and adopted by many practitioners including Google. Thereby, we seek to highlight the novelty of our architecture in incorporating parallel processing to promote throughput. The common architecture will be referred to as *Replication Architecture 1*, or simply as **RA1**, and ours as **RA2**. They are depicted in Figs. 1 and 2 respectively.

In both RA1 and RA2, clients submit transactions directly to only *one* of the n replicas, $R1, R2, \ldots, Rn$; in RA1, there is no parallel processing. More specifically, all replicas first exchange with each other the transactions they directly received, and then order them all identically, prior to processing each of them. That is, all replicas *actively* execute all transactions in the same order. Thus, there are no global conflicts and conflicts encountered are resolved using the same concurrency control (CC) mechanism based on the transactions order.

Barring race conditions, if one replica aborts a given transaction then all would do so. Disagreements due to race conditions and effects of crashes are dealt with during 2PC execution, with each replica acting as the 2PC leader for transactions that it directly received. Spanner [3], Megastore [1] and CockroachDB [16] are recent systems that have adopted RA1. Spanner use TrueTime and others the (leader-based) Paxos or Raft protocols for ordering.

In **RA2** (see Fig. 2), replicas execute the transactions they directly received, in parallel, and using some local concurrency control (CC) mechanism. They then exchange, for each locally survived (i.e., not aborted) transaction, the local transaction identifier and a list of data items accessed and the current value of each write-accessed data, using an order protocol. Global conflicts are detected using data access information and resolved using the order decided on locally survived transactions. Those that survive global conflict resolution proceed to 2PC to be committed. In Fig. 2, $T_1^n, T_3^n, \ldots$ shows a sub-stream of locally survived transactions emerging from R_n, and $T_1^1, T_1^2, T_3^n, \ldots$ show globally survived transactions emerging identically from all replicas.

RA2 is our architecture and uses our ring-based ordering protocol. As noted earlier, any order protocol enables each replica (i) to disseminate its set of locally survived transactions to other replicas, and (ii) to decide an identical order on the combined input super-set. We exploit the former aspect to simplify 2PC implementation by piggybacking relevant information.

Note, however that the traditional 2PC is a leader-based protocol and our order protocol is leaderless. So, 2PC needs to be appropriately adapted. The next two subsections provide the background for this adaptation which is detailed in Sect. 3.1.

2.3 Ring-Based Order Protocol

Our ring-based order Protocol that has been proposed and modelled for performance in [14], is implemented here for $n = 2$ and 3. Figure 3 shows n database replicas being arranged in a logical ring and a *Folder* continually circulating in

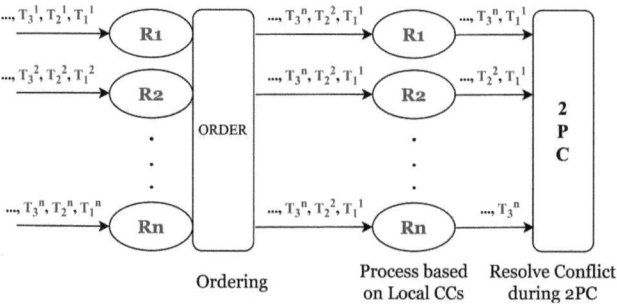

Fig. 1. RA1 - Active Replication: Ordered Processing Everywhere

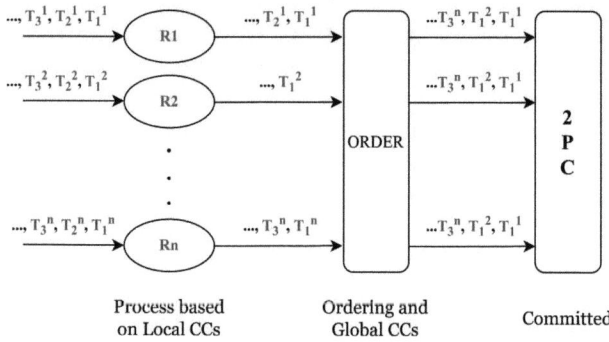

Fig. 2. RA2: Order-based Global Conflict Management before Commit

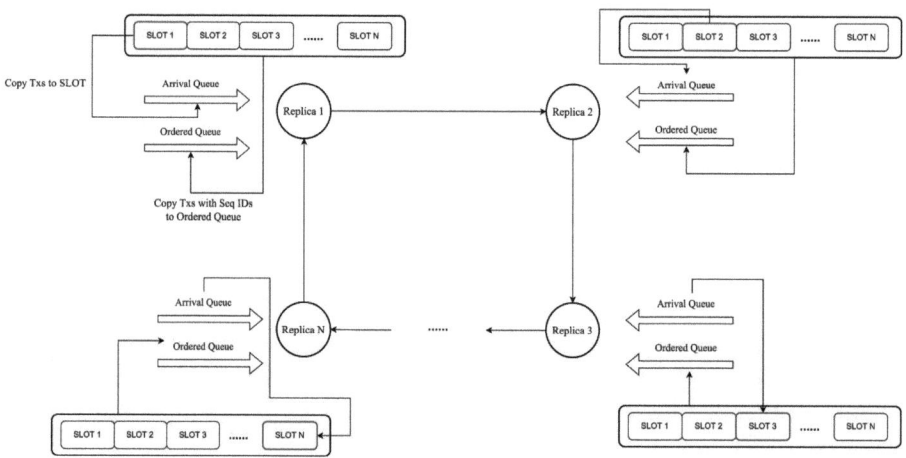

Fig. 3. Ring-Based Order Protocol: Unidirectional Folder Circulation

clockwise direction: moving from Replica R_1 to R_2, R_2 to R_3, ..., R_n to R_1, and so on. The folder contains one slot of fixed size for each replica. Each replica R_i

enqueues each locally-survived and completed transaction in the *Arrival Queue* (see Fig. 3) together with a list of data items accessed and the final values of data items written. When it receives the folder, it enqueues all transactions found in the folder in the *Ordered Queue* as per their sequence number, empties its own slot in the folder, dequeues a transaction from *Arrival Queue* and loads it into its slot with a sequence number that is one more than the largest found in the folder. This loading continues until *Arrival Queue* is empty or there is no more space left in its slot. When it stops, folder is sent to R_{i+1} or R_1 if $i = n$.

2.4 Two-Phase-Commit Protocol (2PC)

2PC [8] protocol involves two phases orchestrated by a lead replica, called the *coordinator* and its execution for any given transaction results in either all replicas committing or aborting that transaction. In the first, *prepare* phase the coordinator sends a *prepare* request to all replicas, asking whether they can successfully commit the transaction. Each replica evaluates commit feasibility by checking for constraint violations and locks write-accessed data items. If a replica can commit, it responds affirmatively; otherwise, negatively. If all respond affirmatively, the coordinator advances to *commit* phase by instructing the replicas to commit the transaction; even if one replica sends a negative or no response (due to crash), the coordinator sends 'abort' to all replicas which abort that transaction. This original 2PC version is adapted in Sect. 3.1 to be executed at the back of the circulating folder in a coordinator-free manner.

3 Ring-Based Ordering and 2PC for 1-Server Abstraction

1-Server abstraction in crash-tolerance literature refers to multiple, crash-prone server replicas 'coming together' to provide a client-level abstraction of a single, crash-free server provided that replica crashes do not exceed a specified threshold. We explain here how this coming-together is realized through our ring-based order protocol using, for simplicity, $n = 2$ replicas of which at most one can crash. Recovering from a crash in a ring structure is discussed in detail in [11], so we focus here on crash-free operation by referring to Fig. 4 where data structures and folder slots of R_1 and R_2 are shown in red and blue respectively.

Replica R_i, $i \in \{1,2\}$, executes the transactions that it directly receives. If a transaction T_j needs to access data item X that has already been accessed by an ongoing, concurrent T_i, then T_j will instantly abort. Transactions that complete their executions without being aborted will have their identifiers together with their *data access lists* queued in the local *Arrival Queue*, AQ_i for short, and also entered in *Local Survived list*, $LS - List_i$ (see Fig. 5). A transaction identifier is the one generated within a replica, concatenated with the replica's id; it is therefore unique within the replicated system, e.g., T_i in R_i can become as $T_i{}^i$ (see also Fig. 2). The *data access list* for $T_i{}^i$, denoted as $DAL(T_i{}^i)$ is the list of all data items that $T_i{}^i$ accessed and the final values of write-accessed data items.

For example, if T_i^i read-accessed X and Z and left the write-accessed Y with value $Y = 10$ on completion, then $DAL(T_i^i)$ will be $\{X, Y = 10, Z\}$.

Whenever R_i receives the circulating, two-slotted folder, it notes down the largest sequence number in the folder and copies all entries in each slot of the folder into its *Ordered Queue*, OQ_i for short, as per their sequence number. It then empties its own slot in the folder and then loads as many AQ_i entries into its slot, with each loaded entry assigned a sequence number continuing sequentially from the largest noted. If, say, the latter is 110 and R_i loads nine entries, their sequence numbers would be $111, 112, \ldots, 119$. Once its slot is loaded to its capacity, the folder is sent to the next replica in the ring, $R_j, j \neq i$; when R_j receives the folder, it would note the largest sequence number as 119.

Two remarks are in order. First, when R_i is in possession of the folder, it does not modify the contents of any slot other than its own. Secondly, contents of all slots of the arriving folder are copied for ordering, including its own slot which was loaded when R_i had the folder in the previous cycle; those being loaded now would be copied and entered into OQ_i when the folder returns next. It is easy to see that replicas enter all locally-survived transactions into their respective OQ in the same (sequence number based) order.

Whenever OQ_i is non-empty, R_i dequeues the first item and compares the DAL in the dequeued item with the DAL of *every* entry stored in the *Global-Survived List*, $GS - List_i$, that it maintains (see Fig. 5). If a conflict is detected with any $GS - List_i$ entry, the dequeued item is entered in *Global-Aborted List*, $GA - List_i$ and the corresponding transaction must be aborted during 2PC; otherwise, the dequeued item is entered in $GS - List_i$ for commit during 2PC.

Let us consider an example: let $\{T_i^i, DAL(T_i^i)\}$ be the item freshly dequeued from OQ_i and let $\{T_h^j, DAL(T_h^j)\}$ be some entry in $GS - List_i$ which would refer to transaction T_h that locally survived in R_j and was sequenced before T_i^i and hence is already in $GS - List_i$. If $DAL(T_h^j)$ indicates that T_h^j wrote data item X when it was executed at R_j and $DAL(T_i^i)$ indicates that T_i^i read data item X when it was executed in parallel at R_i, then it would be treated as a global *write-read* conflict and the later ordered T_i^i is marked for abort.

All replicas will reach the same outcome while checking a given item dequeued from their respective OQ *provided* that their respective GS-Lists also have identical contents at the time of their checking. The latter cannot be guaranteed as globally-survived transactions that have gone through the next stage 2PC execution must be removed from GS-List. Due to inherent asynchrony in distributed computing, replicas may decide differently: for example, R_i entering $\{T_i^i, DAL(T_i^i)\}$ in $GS - List_i$ and R_j entering the same $\{T_i^i, DAL(T_i^i)\}$ in $GA - List_j$ instead. This can occur, for example, if T_h^j had been removed from $GS - List_i$ following its commit in R_i but is yet to be removed from $GS - List_j$ at the time when R_i and R_j dequeued $\{T_i^i, DAL(T_i^i)\}$ from their respective GS-List. This inconsistency will be sorted out during 2PC when R_j would respond negatively for committing T_i^i and force R_i also to abort T_i^i. Thus, all replicas either commit or abort a given input transaction irrespective which replica directly executed that transaction; i.e., 1-server abstraction is ensured.

3.1 2PC on the Back of Ring-Based Ordering

To implement 2PC using the circulating folder, we require that the folder, in addition to n slots, contains a list of *Blocks*, one *block* for each *transaction* that has entered the *GS-List* and hence is ready to be committed through 2PC execution. As noted in Subsect. 2.4, all replicas must respond affirmatively after *prepare* phase for a commit outcome and a replica's 'no' acts as a 'veto'.

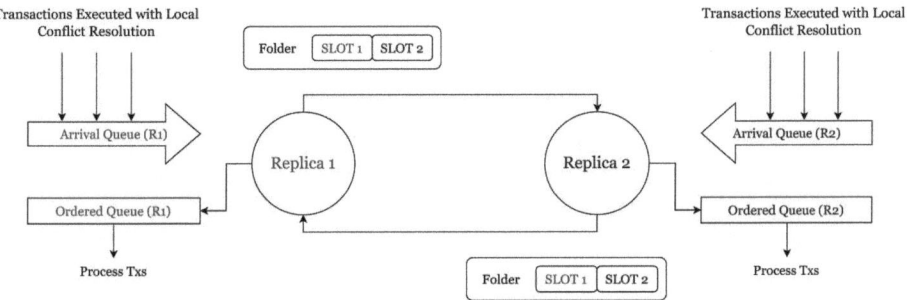

Fig. 4. Ordering Locally-Survived Transactions Using Circulating Folder.

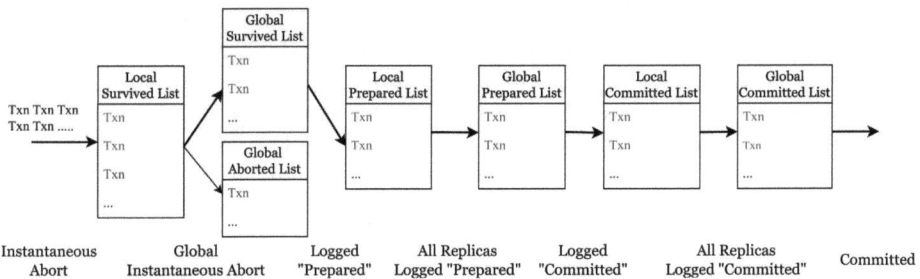

Fig. 5. Lists for Transaction Lifecycle Management

The structure of a transaction's *block* reflects this information collection: transaction (global) identifier (e.g., T_h^j) accompanied by an integer vector of n indices - one for each replica (see Fig. 5 for $n = 3$). The vector in the block for transaction T is denoted as V_T. $V_T[i], 1 \leq i \leq n$, is 0, 1 and 2 respectively implies that R_i possibly started 2PC for T and is *preparing*, R_i completed *prepare* phase and is ready for *commit* phase, and R_i committed T; $V_T[i] = -1$ indicates that R_i responded negatively for T in *prepare* phase and therefore $V_T[j]$ can never become 2 for any replica R_j.

Each R_i also maintains four more lists (shown in Fig. 5) and their abbreviated names follow this intuitive convention: **S** for Survived, **L** for Local, G for Global, and **P, C** and **A** for Prepared, Committed and Aborted respectively.

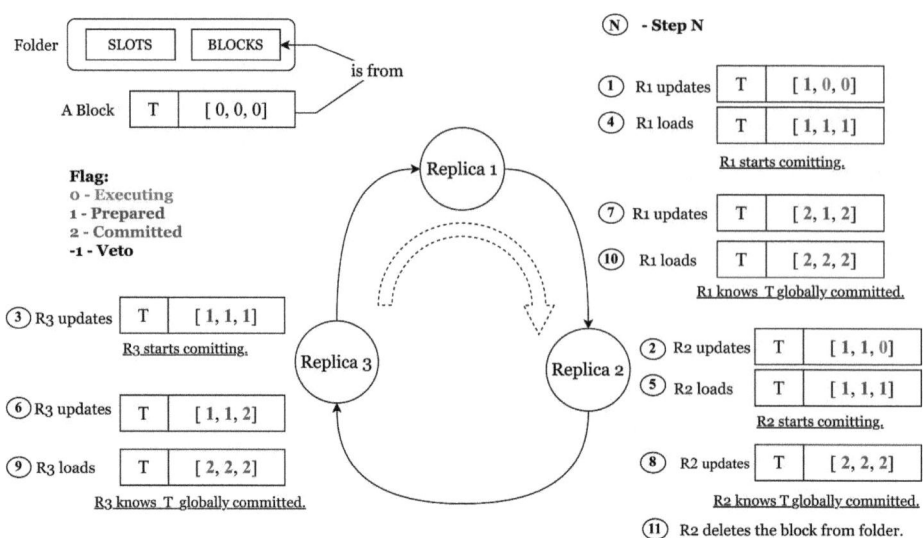

Fig. 6. Ring-Based 2PC Steps in Committing Transaction T

For space reasons, we explain the workings of our *Ring-Based 2PC* for transaction T by referring to Fig. 6 where it is assumed that all replicas enter T in their respective *GS-List*. The other case of only some replicas doing so and others entering T in *GA-List* is simpler and discussed next. We also assume that a replica optimistically starts preparing for committing T (i.e., the *prepare* phase) as soon as it enters T in its *GS-List* and similar optimism is common in 2PC implementations; *Replica i*, $1 \leq i \leq 3$, in Fig. 6 is simply referred to as R_i.

The description based on Fig. 6 involves 11 Steps; transition to the next step corresponds to a given replica receiving the folder circulating in the ring.

Step 1: We assume that *Replica 1* (R_1) is the first replica to notice that the folder does not yet have a block for T that is in its $GS - List_1$. So, it adds a block for T in the *Blocks* part of the received folder with $V_T = [0,0,0]$. Suppose that preparation for T is locally completed; $V_T =$ is set to [1,0,0] and the entry for T in $GS - List_1$ is moved to *Locally Prepared List*, $LP - List_1$. (If not completed, $V_T[1]$ is left unchanged at 0 and these operations are to be carried out at the earliest instance when R_1 receives the folder after it has completed *prepare* for T.) When R_1 is done with the folder, it transmits the folder to R_2.

Step 2: Suppose that R_2 has also completed *prepare* for T when it receives the folder. It sets $V_T =$ to [1,1,0] and moves the entry for T from $GS - List_2$ to $LP - List_2$. When ready, R_2 will transmit the folder to R_3.

Step 3: Suppose also that R_3 also completed *prepare* phase for T when it receives the folder (from R_2). It will behave like *Replica 2* stated in Step 2, except that $V_T = [1,1,1]$ and R_3 now deduces that all 3 replicas have completed *prepare* for T, starts the next 2PC phase *commit* for T and moves the entry for T from $LP - List_3$ to *Global-Prepared List*, $GP - List_3$.

Step 4: When R_1 receives the folder with $V_T = [1,1,1]$, it deduces that all replicas have done *commit* for T; so, it moves the entry for T from $GP - List_1$ into $GP - List_1$, *Global-Prepared List* and starts *commit* for T.

Step 5: R_2, on receiving the folder, starts *commit* for T, like R_1 in Step 4.

Step 6: Recall that R_3 started *commit* for T in Step 3. It is possible, and we here assume so, that when R_3 receives the folder now, it has completed *commit* for T; if so, it will move the entry for T from $GP - List_3$ to $LC - List_3$ and sets $V_T = [1, 1, 2]$. (If *commit* is not completed, these actions will be done at the earliest instance when R_3 receives the folder after *commit* for T is locally completed.)

Step 7: Suppose that R_1 also completed *commit* for T when it receives the folder now. It sets $V_T = [2, 1, 2]$ and moves entry for T from $GP - List_1$ to $LC - List_1$, the *Local Committed List* (see Fig. 5).

Step 8: Retaining the assumption that a replica completes *commit* for T when the folder makes one round in the ring, R_2, when it receives the folder, sets $V_T = [2, 2, 2]$ and moves entry for T from $GP - List_2$ to $LC - List_2$. Additionally, it deduces from $V_T = [2, 2, 2]$ that T is committed everywhere and moves entry for T from $LC - List_2$ to $GC - List_2$.

Step 9: R_3 locally carries out the additional operations done by R_2 in Step 8.

Step 10: R_1 locally carries out the operations done by R_3 in Step 9. Note: all replicas now have T in their $GC - List$.

Step 11: *Garbage Collection Rules* are as follows. Whenever R_i receives the folder with V_T already set to $[2, 2, 2]$ for any T, it deletes the entry for T in $GC - List_i$ if it *already* exists in $GC - List_i$; otherwise, it deletes the block for T in the *Blocks* part of the incoming folder. So, R_2 discards its $GC - List_2$ entry for T here when folder arrives with $V_T = [2, 2, 2]$.

Continuing on, R_3 and R_1 will discard their $GC - List$ entry for T in Step 12 and 13 respectively, and R_2 will discard the *Blocks* entry for T in Step 14.

Executions with $T \in GA - List_i$ for some or all $R_i, 1 \leq i \leq 3$. Suppose that R_1 has entered T in its $GA - List_1$ and receives the folder in Step 1 above with no *Blocks* entry for T. It will create an entry with $V_T = [-1, 0, 0]$ Any replica R_j that receives the folder sees this veto for T will be in one of *two* situations: (i) it also has T in $GA - List_j$ in which case it sets $V_T[j]$ to -1, or (ii) it has T in $GS - List_j$ in which case it aborts any *prepare* done or being done for T, moves the $GS - List_j$ entry for T into $GA - List_j$ and sets $V_T[j] = -1$.

Another scenario of interest is that R_1 has entered T in its $GS - List_1$ and behaves as in Step 1 above, and only replica that has entered T in its $GA - List$ is R_2 and/or R_3. Let us assume that it is only R_2 with T in $GA - List_2$. In Step 2, R_2 will execute the actions of (i) above and R_3 and R_1 will execute the actions of (ii) above in Step 2 and 3 respectively. *Garbage Collection Rules* are as before except that the incoming folder should have V_T already set to $[-1, -1, -1]$ (instead of $[2, 2, 2]$) and $GA - List$ is used instead of $GC - List$. Thus, R_2, R_3

and R_1 will discard their $GA - List$ entry for T in Step 5, 6 and 7 respectively; in Step 8, R_2 will discard the *Blocks* entry for T in the received folder.

4 Implementation and Performance Evaluation

We implemented the ring-based order protocol with $n, n = 2, 3$, servers forming the ring and measured its performance in terms of *Latency* that is defined as the time elapsed between the instance a locally-survived transaction enters the *Arrival Queue (AQ)* of a replica and the moment it has been entered in the *Ordered Queue* of all replicas (see Fig 3), the given slot size k is 1KB in the *Folder* for every replica. *Latency* thus includes the wait-time in AQ and the time spent in the circulating folder.

We compared the uniform *Measured Latency* with the *Estimated Latency* from model-based analytical approximations presented in [14]. While the former is measured by repeating experiments on the implemented system, the latter requires measuring two system parameters: the average time (α) taken by replicas to process a received folder and the average folder transmission time (β) between replicas on the ring; measurements found $\alpha = 10^{-3}$ secs and $\beta = 10^{-5}$ secs.

Analysis in [14] states that the system is stable when the average arrival rate into AQ, denoted as λ, is: $\lambda < \frac{k}{n^2 k\alpha + \beta}$, where k is the number of slots in the folder which is n. Since β is negligibly small compared to $n^2 k\alpha$, we can conclude that the largest stable λ is inversely proportional to n^2 which will guide our choice of λ values for experiments. Latency is estimated as follows.

For any stable λ, a unique solution for $s \in (0, 1)$ exists:

$$s = \frac{\lambda(n\alpha + \beta)}{1 - \lambda n(n-1)\alpha} . \tag{1}$$

The average queue size L at each replica is estimated as $\frac{s}{1-s}$ and the estimated Latency (by Little's Law) is $\frac{L}{\lambda}$.

4.1 Two Replica System

With $n = 2$, the maximum stable λ is to be no larger than 249. We began by measuring the Latency under varying arrival rates, specifically at $\lambda = 50$, 100, 150, 200, and 249. Starting from $\lambda = 200$, a notable increase in both the measured and estimated L values is observed, indicating a sharp transition in system behaviour. At $\lambda = 249$, the values become extremely large, reflecting system is approaching instability or saturation.

We conducted additional experiments focusing on the critical range $239 \leq \lambda \leq 249$, collecting a more granular set of data to understand the system dynamics in this region better. Detailed experimental results for $239 \leq \lambda \leq 249$, including the estimated latency, are summarized in Table 1.

Our experimental results reveal that the system achieves its optimal performance when the transaction arrival rate λ lies under 247. When λ approaches

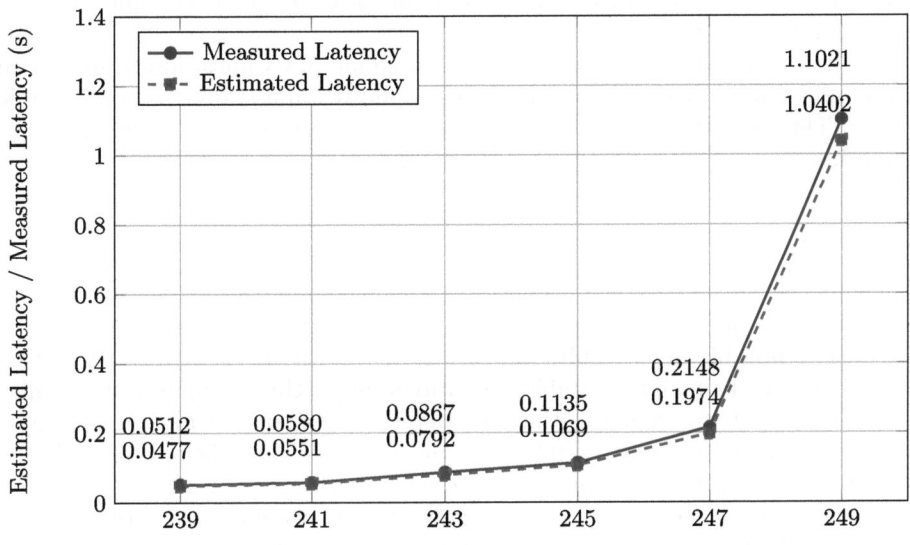

Fig. 7. Two Replicas with Increasing λ. $\lambda \in [239, 249]$

Table 1. Two Replicas: Results for λ in the Range 239 to 249

λ	Measured L	Estimated L	Estimated Latency (s)	Measured Latency (s)
239	11.3890	11.5451	0.0477	0.0512
241	13.2804	14.4213	0.0551	0.0580
243	19.2459	19.1017	0.0792	0.0867
245	26.2024	28.0598	0.1069	0.1135
247	48.7488	52.0955	0.1974	0.2148
249	259.0055	331.4503	1.0402	1.1021

Table 2. Three Replicas: Results for λ in the Range 20 to 110

λ	Measured L	Estimated L	Estimated Latency (s)	Measured Latency (s)
20	0.0712	0.0734	3.56×10^{-3}	3.73×10^{-3}
40	0.1774	0.1882	4.435×10^{-3}	4.521×10^{-3}
60	0.3686	0.3931	6.143×10^{-3}	6.495×10^{-3}
80	0.8119	0.8625	1.015×10^{-2}	1.105×10^{-2}
100	2.8382	3.04	2.838×10^{-2}	2.897×10^{-2}
110	32.6624	37.2022	0.2969	0.3175

248 or 249, the system performance degrades noticeably but remains operational. However, when λ exceeds 249, the system surpasses its capacity limits, leading to system instability.

Finally, as illustrated in Fig. 7, the measured latency values closely align with estimated ones, indicating that the system operates in a stable and predictable regime. This consistency confirms that $\lambda \leq 247$ represents a robust region for maintaining high throughput with bounded latency.

4.2 Three Replicas: $n = 3$

Following the same analytical framework, when the number of replicas increases to $n = 3$, the system's stability threshold for the maximum permissible value of λ decreases notably. The results indicate that λ must be constrained to be no larger than 110 to maintain stable operations under this configuration. Similar to the case with two replicas, the parameter k continues to have a negligible impact on system stability, reinforcing the robustness of the protocol design against variations in transaction size or queue depth at this scale.

Table 2 shows the experimental results for a system with three replicas ($n = 3$), evaluated under varying transaction arrival rates λ ranging from 20 to 110. The Table presents the measured and estimated values of L, representing the average number of transactions in the system, as well as the corresponding measured latency and measured delay values.

The data shows that the system load increases correspondingly as λ increases. The measured and estimated L values are closely aligned across all tested λ values, indicating that the analytical model accurately estimates the average system load under varying traffic intensities.

In terms of latency performance, the estimated latency and measured latency also exhibit a strong correlation. Both metrics increase as λ grows, which is consistent with queuing theory expectations. For lower λ values (e.g., $\lambda = 20$), the system maintains very low latency, with delays in the millisecond range. As the transaction arrival rate increases towards $\lambda = 110$, both latency and delay reach sub-second values, highlighting the impact of increased load on system responsiveness.

As λ increases, the system load and message delay also increase predictably, following expected queuing behaviour. The close alignment between theoretical predictions and empirical results validates the accuracy and robustness of the proposed analytical model. Even under higher traffic conditions, the system maintained consistent and stable behavior, suggesting that the message ordering protocol scales effectively in distributed environments with multiple replicas.

5 Conclusion

The Ring Based Ordering and 2PC Protocols provide an effective solution for database replication while ensuring the most desirable ACID properties which eliminate state divergence and the need for complex reconciliation mechanisms.

Moreover, by allowing parallel processing at replicas, a higher throughput is maintained. Thus, our ring based replication architecture represents a significant step towards high-performance, fault-tolerant database replication, providing a novel, robust foundation for modern distributed systems.

Our experiments on the implemented system show that actual performance is remarkably close to our analytical estimations. An advantage of this closeness is that we can use our model-based estimations to adapt the system for prevailing arrival rates, e.g., by increasing the number of slots per replica. We are currently designing order protocols with multiple circulating folders. Our initial assessments show that even higher throughputs are possible but at the cost of increased latency. Thus, the single-folder protocol presented here represents one end of a range of ring-based options that can be exercised.

References

1. Baker, J., et al.: Megastore: providing scalable, highly available storage for interactive services. In: Proceedings of the Conference on Innovative Data system Research (CIDR), pp. 223–234 (2011)
2. Chandra, T.D., et al.: Unreliable failure detectors for reliable distributed systems. J. ACM (JACM) **43**(2), 225–267 (1996)
3. Corbett, J.C., et al.: Spanner: Google's globally distributed database. ACM Trans. Comput. Syst. **31**(3) (2013)
4. Durner, D., Neumann, T.: No false negatives: accepting all useful schedules in a fast serializable many-core system. In: 2019 IEEE 35th International Conference on Data Engineering (ICDE), pp. 734–745. IEEE (2019)
5. Ezhilchelvan, P., Mitrani, I., Webber, J.: On the degradation of distributed graph databases with eventual consistency. In: Computer Performance Engineering, pp. 1–13. Springer (2018)
6. Ezhilchelvan, P., Mitrani, I., Webber, J.: Modeling the gradual degradation of eventually-consistent distributed graph databases. Queueing Models Serv. Manag. **3**(2), 235–253 (2020)
7. Ezhilchelvan, P., Mitrani, I., Webber, J., Wang, Y.: Evaluating the performance impact of no-wait approach to resolving write conflicts in databases. In: European Workshop on Performance Engineering, pp. 171–185. Springer (2023)
8. Gray, J.N.: Notes on data base operating systems. Oper. Syst. Adv. Course 393–481 (2005)
9. Gray, J.: Notes on data base operating systems. In: Operating Systems, An Advanced Course, pp. 393–481, Berlin, Heidelberg (1978). Springer-Verlag
10. Gray, J., Lamport, L.: Consensus on transaction commit. ACM Trans. Database Syst. (TODS) **31**(1), 133–160 (2006)
11. Guerraoui, R., et al.: Throughput optimal total order broadcast for cluster environments. ACM Trans. Comput. Syst. (TOCS) **28**(2), 1–32 (2010)
12. Kalavadia, B., et al.: Adaptive partitioning using partial replication for sensor data. In: Distributed Computing and Internet Technology: 15th International Conference, pp. 260–269. Springer (2019)
13. Kemme, B., Jiménez-Peris, R., Patiño-Martínez, M.: Database Replication. Synthesis lectures on data management. Morgan & Claypool Publishers (2010)

14. Liu, Y., Ezhilchelvan, P., Mitrani, I.: Design and analysis of distributed message ordering over a unidirectional logical ring. In: European Workshop on Performance Engineering, pp. 1–13. Springer (2024)
15. Melton, J.: ANSI/ISO SQL-92 specification. Online (1994)
16. Taft, R., et al.: CockroachDB: the resilient geo-distributed SQL database. In: Proceedings of the 2020 ACM SIGMOD International Conference on Management of Data, SIGMOD '20, pp. 1493–1509, New York, NY, USA (2020). Association for Computing Machinery
17. Wiesmann, M., et al.: Database replication techniques: a three parameter classification, pp. 206–215 (2000)
18. Wiesmann, M., Schiper, A.: Comparison of database replication techniques based on total order broadcast. IEEE Trans. Knowl. Data Eng. **17**(4), 551–566 (2005)

Blockchain-Backed Fuzzy Search for Semi-structured Translation Data: A Scalable Hybrid Approach with Hyperledger Fabric and Elasticsearch

Edvan Soares(✉) and Valeria Times

Computer Center, Federal University of Pernambuco, Recife, Pernambuco 50740560, Brazil
{ejsj2,vct}@cin.ufpe.br

Abstract. Translation Memory (TM) systems are critical components of modern computer-aided translation (CAT) workflows. However, centralized TM platforms often lack transparency, user control, and verifiable trust guarantees. This paper introduces a decentralized architecture that enables scalable fuzzy search over TM segments while ensuring data provenance and integrity through blockchain validation. The proposed solution integrates Elasticsearch for high-performance approximate matching with Hyperledger Fabric as a trust-enforcing validation layer. The proposed system is designed to be interoperable with standard CAT tools via a backend gateway that performs fuzzy retrieval and verifies match authenticity using smart contracts. Importantly, only hashed metadata is stored on-chain, preserving confidentiality while enabling auditability. We conducted four experimental rounds with datasets of 100k, 500k, 1M and 10M segments to assess the system's performance and scalability. Results show that the architecture maintains sub-second query times even at scale, with the blockchain validation layer introducing minimal overhead. These findings demonstrate the feasibility of integrating decentralized trust mechanisms into real-time linguistic data systems. The research illustrates how database engineering principles can be effectively combined with blockchain technologies to meet the evolving demands of secure and decentralized collaboration.

Keywords: Fuzzy Search · Blockchain · Translation Memory · Data Privacy · Distributed Systems

1 Introduction

The translation industry relies on translation memory (TM) systems to reuse previous translations, enhancing consistency and productivity [13]. Therefore, the increasingly collaborative nature of these systems creates new challenges regarding data ownership, privacy, and trust issues, where organizations require better control of their linguistic data while needing assurance about its integrity and flexible search capabilities that scale efficiently.

Fuzzy search mechanisms are a cornerstone of modern TM systems, allowing users to retrieve approximate matches based on linguistic similarity. Implementing these mechanisms within decentralized infrastructures introduces non-trivial technical trade-offs, where systems must simultaneously support low-latency retrieval, strong privacy guarantees, and tamper-proof auditing, all while maintaining compatibility with human translators' fuzzy matching expectations.

Traditional TM systems operate under centralized paradigms that limit transparency, auditability, and multi-party trust. Attempts to decentralize these systems often ignore efficient retrieval performance or require renouncing control to third-party search providers. Therefore, this paper addresses the following research problem: *How can we design a decentralized architecture that enables scalable fuzzy search over translation segments while ensuring data integrity, traceability and privacy?*

To address this challenge, we propose a hybrid decentralized system architecture integrating Hyperledger Fabric—a permissioned blockchain platform—with Elasticsearch, a distributed search engine known for its efficient fuzzy matching capabilities. The proposed architecture enables CAT tools to interact with a gateway application that performs high-performance search operations and validates results via blockchain-based smart contracts. The main contributions of this research are as follows:

- A novel system architecture for enabling decentralized fuzzy searches over semi-structured translation data;
- A functional implementation that integrates Elasticsearch for data retrieval and Hyperledger Fabric for on-chain verification;
- An experimental evaluation measuring retrieval latency and system scalability.

This research follows the Design Science Research (DSR) methodology, which structures the investigation into a series of iterative and systematic phases: problem identification, objective definition, design and development of the artifact, demonstration, evaluation, and communication [11]. This approach is well-suited for addressing socio-technical problems by creating and assessing innovative IT artifacts.

The paper has the following structure. Section 2 presents some related work, Sect. 3 presents the problem identification and motivation, explaining the reason for the proposed solution. Section 4 outlines the solution objectives and design requirements. Section 5 explains the proposed architecture design and development process, including its components and data flow. Section 6 illustrates the artifact demonstration by showing its application in a real-world Translation Memories use case. Section 7 demonstrates the evaluation results from experimental tests conducted. Section 8 explores the approach's broader implications, generalization, and limitations. The work's final section presents conclusions and future research directions in Sect. 9.

2 Related Work

Research on approximate string matching and fuzzy retrieval is well-established within the information retrieval (IR) community. Classic approaches rely on edit distance algorithms such as Levenshtein distance, token-based similarity, or n-gram matching [17]. More recent studies explore neural fuzzy retrieval methods that leverage translation memory context to improve semantic matching and translation quality [4,6]. Modern search engines like Elasticsearch implement some of those techniques over scalable indexing structures [8]. However, these tools are typically centralized and do not provide auditability or verifiable trust mechanisms.

On the other hand, blockchain technologies have been adopted to guarantee data integrity and track access in distributed environments through their implementation. Hyperledger Fabric is a permissioned blockchain framework that enterprises prefer because it offers modularity and supports smart contracts and identity management features [1]. Research on blockchain verification methods has been explored in different domains, such as healthcare and secure document exchange systems [2].

Feng et al. [5] present a blockchain-enabled framework for privacy-preserving data sharing that incorporates fuzzy search functionality over encrypted content. Their approach is tailored to secure document access, emphasizing confidentiality and access control through blockchain-based audit trails. While the combination of fuzzy search and blockchain is conceptually aligned with our proposal, their solution is not optimized for performance in high-volume scenarios, nor does it account for the linguistic variability present in translation memory segments. In contrast, our system prioritizes efficiency and interoperability, leveraging Elasticsearch for fast approximate retrieval over semi-structured language data and Hyperledger Fabric for decentralized validation. Rather than focusing on encrypted search, our architecture addresses real-world demands of collaborative TM usage in production CAT tool ecosystems.

Zou et al. [18] introduce a fuzzy keyword searchable encryption scheme based on blockchain that supports multilevel access control and on-chain validation. Although their model improves security and trust in encrypted search scenarios, it focuses primarily on keyword fuzziness embedded in encryption, and it does not support large-scale approximate matching over unstructured text. Our architecture addresses these limitations by separating the retrieval and validation layers, employing Elasticsearch to handle high-throughput fuzzy queries and Hyperledger Fabric to guarantee result integrity, thus supporting real-time TM applications at scale.

As far as we know no existing work combines scalable fuzzy search with blockchain-based verification in the context of semi-structured translation data. While recent efforts have improved the integration of translation memories into neural architectures [4,15], none of them offer a modular system supporting large-scale fuzzy search with on-chain validation. Our architecture addresses this gap by offering a modular, interoperable solution that bridges performance and trust in privacy-sensitive environments, dealing with challenges to efficient data sharing using blockchain [9].

3 Problem Identification and Motivation

The computer-aided translation (CAT) ecosystem relies heavily on Translation Memories (TMs) as essential components. The system functions to store and retrieve source-target text segments that are already aligned for future use. The stored segments enable translators to preserve document consistency while decreasing manual work by suggesting already translated content. The query process of TMs depends on fuzzy search algorithms, which measure the linguistic similarity between the translator input and stored segments [13]. The utility of TMs depends on their ability to perform approximate matching while handling linguistic variations and delivering results within interactive timeframes.

Traditional TM systems are typically deployed in centralized architectures controlled by commercial platforms or enterprise servers [12]. While these setups benefit from managed infrastructure and controlled access, they also introduce critical limitations. The centralized management of TMs creates significant risks because it compromises data ownership rights and transparency while enabling the unauthorized use of linguistic assets. The data contributors to TMs, including translators and organizations, maintain minimal control over their data utilization, distribution, and financial exploitation. The unequal power dynamics between parties create trust issues that prevent institutions from working together, especially when intellectual property, legal compliance, and confidentiality requirements are essential.

The growing globalization of translation services and distributed collaboration models requires TM systems to offer decentralized functionality, user control, and robust privacy protection. Users in these environments require guarantees that their data remains secure from unauthorized access, tampering, and misuse. The solution must maintain the high responsiveness that CAT tools need because they depend on fast fuzzy matching for human-in-the-loop translation workflows.

The implementation of decentralized TMs requires addressing multiple significant technical challenges. The technical advantages of blockchain platforms include powerful immutability features, provenance tracking, and auditable access control capabilities. The system requires these features to establish trust between multiple parties while eliminating the need for central authority control. However, the performance of blockchain platforms deteriorates significantly when they handle high-volume fuzzy search operations and complex or latency-sensitive retrieval queries [1]. On the other hand, the search engine Elasticsearch provides fast approximate matching and large-scale indexing capabilities. However, it does not have built-in validation mechanisms, trust enforcement, or access accountability features [8].

This architectural gap between trust and performance creates a fundamental obstacle: how can we simultaneously provide the efficient fuzzy retrieval demanded by translation professionals and the verifiability and traceability required in decentralized data ecosystems? To overcome this gap, a new architectural approach is needed to separate retrieval from verification, allowing both to operate harmoniously. Without such an approach, retrieval efficiency or data trustworthiness must be compromised—an unacceptable outcome in modern translation infrastructures where both are critical.

The challenge goes beyond the translation industry. Many applications deal with semi-structured textual data that must be queried approximately and securely across organizational or jurisdictional boundaries. Examples include legal archives, biomedical research repositories, multilingual knowledge bases, and collaborative annotation systems [7]. These domains face similar challenges because they need to extract value from approximate matches while maintaining data integrity, provenance, and user-defined access policies. Therefore, solving this problem within the TM domain offers a blueprint for building trustworthy, high-performance information retrieval systems across a broad spectrum of decentralized knowledge infrastructures.

4 Objectives of the Proposed Solution

The research establishes a decentralized system architecture that enables efficient and reliable fuzzy search operations throughout TM segments. The objective emerges from the need to match linguistic retrieval performance requirements with distributed collaborative data integrity, privacy, and trust guarantees. In line with the Design Science Research methodology, this solution is not limited to demonstrating technical feasibility; it must also meet the organizational and social requirements inherent in professional translation workflows.

The system architecture needs to perform fast approximate retrieval of translation segments from large databases containing millions of entries. The system requires a real-time fuzzy matching search engine like Elasticsearch, which enables customizable similarity metrics and efficient indexing strategies. The system requires more than retrieval performance in distributed contexts. The system architecture must eliminate centralized control of translation memories through a decentralized trust layer that verifies segment authenticity without degrading system performance.

To address these requirements, the solution integrates a blockchain platform as a verification layer that stores cryptographic proofs and metadata associated with each registered segment. By doing so, the system guarantees that retrieved results are relevant from a linguistic standpoint and verifiable in terms of authorship, timestamp, and integrity. The proposed model differs from standard TM systems because it allows users and organizations to maintain data control and monitor all system interactions without depending on centralized servers or opaque platforms.

The system also requires interoperability as one of its essential goals. The architecture needs to integrate with current CAT tools through standard communication interfaces for seamless integration. The system design enables smooth integration with existing CAT tools through RESTful APIs, which prevents disruptions to translator workflows and boosts adoption potential in professional settings. Additionally, the system must support modular deployment because organizations should be able to host and manage their components (e.g., instances of search engines, blockchain peers) based on their privacy and infrastructure needs.

In practical terms, the architecture must fulfill several functional and non-functional requirements. Functionally, it should allow CAT tools to issue secure search requests, retrieve candidate matches based on fuzzy similarity scores, and validate each match through smart contract logic. Non-functionally, the system must sustain sub-second query response times, ensure secure and authenticated communication between all components, and offer scalability without compromising privacy guarantees. These combined goals define the foundation upon which the system was designed and implemented, and they directly influence the technical decisions presented in the following sections.

5 Design and Development

The system architecture proposal combined real-time fuzzy search performance requirements with blockchain technology guarantees of trust, data integrity, and decentralization. The system uses Elasticsearch for high-throughput retrieval while Hyperledger Fabric handles verification, auditing, and data traceability. The system design separates responsibilities to achieve scalability and trust enforcement without compromising usability or performance.

At the center of the architecture is a gateway service, which acts as an intermediary between CAT tools and the backend components. The CAT tool—representing the translator's interface—sends fuzzy search requests to this gateway. These requests are then forwarded to Elasticsearch, which performs approximate string matching over an indexed corpus of translation memory segments. The matching algorithm is based on similarity metrics such as Levenshtein distance [16], allowing for flexible and linguistically relevant retrieval even when exact matches are absent.

Figure 1 illustrates the overall architecture and the interaction between its components, which include: (1) the CAT tool client, responsible for initiating search queries; (2) the gateway service, which coordinates search and validation; (3) Elasticsearch, which provides real-time fuzzy matching; and (4) Hyperledger Fabric, which stores tamper-evident metadata and executes smart contracts for validation.

The architecture implements essential mechanisms that protect privacy and security while enabling interoperability. The CAT tool communicates with the gateway through HTTPS encryption while API keys control access to the system. The system protects translation content confidentiality by storing hashed segment representations on-chain instead of the original text. The Hyperledger Fabric network maintains strict membership policies restricting peer operation and transaction submission to authorized organizations. The ledger records every

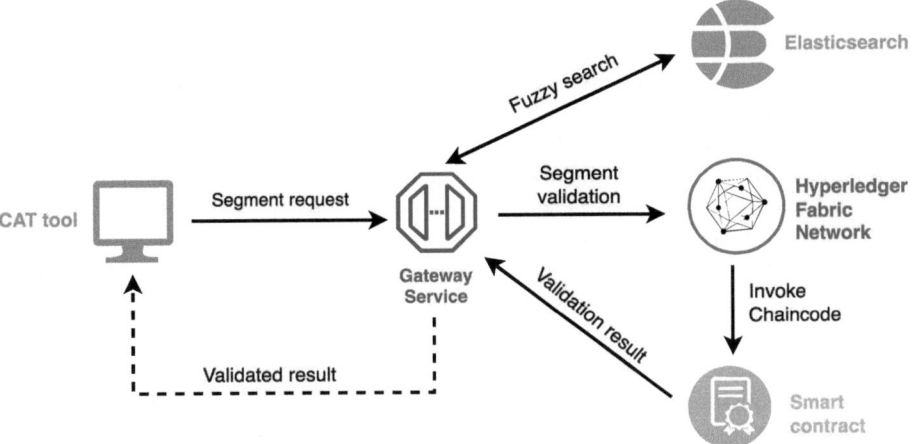

Fig. 1. Architecture integrating fuzzy search (Elasticsearch) with blockchain-based validation (Hyperledger Fabric).

validation request, which provides complete audit capabilities for tracking access and verification actions.

This design allows organizations to deploy modules across different operational environments. The system enables organizations to operate their components, including private Elasticsearch nodes and Fabric peers, through their own infrastructure instead of depending on centralized systems. Thus, the system is positioned as a privacy-aware, scalable, and verifiable platform for decentralized translation memory search, with broader applicability to other domains that require secure fuzzy retrieval over semi-structured textual data.

6 Demonstration

A fully functional prototype was developed and deployed in a controlled cloud-based environment to assess the feasibility and operational effectiveness of the proposed architecture in a realistic scenario. This demonstration aimed to validate whether the architectural components could work cohesively to support fuzzy TM retrieval with blockchain-backed validation while maintaining low-latency performance and system reliability under concurrent query loads.

The implemented system included all major components described in the architectural model (Fig. 1): (i) a computer-assisted translation (CAT) tool emulator acting as the user interface; (ii) a backend gateway service responsible for query orchestration and validation; (iii) an Elasticsearch node configured for high-performance fuzzy matching using Levenshtein-based similarity; and (iv) a permissioned Hyperledger Fabric network serving as the blockchain validation layer. These components were designed to operate modularly, simulating a real-world deployment scenario for enterprise-grade translation workflows.

The deployment used Docker containers to achieve environment isolation, simplify orchestration, and ensure the reproducibility of all services, hosted in a Linux-based virtual machine with 32 GB of RAM and eight virtual CPUs, providing adequate resources to replicate concurrent access patterns without performance bottlenecks. The CAT tool emulator issued parallel search queries, each containing a source translation segment. These requests triggered the complete processing flow, starting with fuzzy matching over the indexed segments in Elasticsearch and proceeding to validation via smart contracts deployed on Hyperledger Fabric. During this process, the gateway transformed each retrieved segment into a cryptographic hash, which was submitted to the blockchain to be validated by a smart contract that verifies the segment's existence. Once validated, the gateway retrieves the associated metadata, including contributor identifiers, timestamps, and data provenance indicators, and returns to the CAT tool emulator, completing the full query lifecycle. Communication between components was encrypted using HTTPS, and the gateway enforced access control via API keys to simulate secure multi-user environments.

The demonstration showed how the architecture worked as an interoperable system. Despite using heterogeneous technologies, the system operated without errors or performance issues while maintaining smooth interactions between the search engine and blockchain. The gateway's API-centric design allowed the system architecture to connect with existing CAT tools through APIs without changing anything on the client side. This is critical for adoption in enterprise translation environments, where existing tools and workflows must be preserved. Furthermore, the demonstration showed how the system maintained stability and responded well to parallel execution. The system processed one hundred search requests at once across test datasets, which proved its ability to scale for production needs. The results confirmed the architectural prediction that blockchain validation could operate independently from retrieval logic, resulting in minimal latency that did not impact user experience.

Overall, the demonstration phase confirmed that the proposed hybrid architecture is theoretically sound and practically viable. It reinforced the claim that decentralized validation and high-performance fuzzy retrieval coexist within the same system. It laid the groundwork for further deployment and evaluation in real-world CAT environments.

7 Evaluation

Following the successful demonstration of the system, we conducted a structured performance evaluation to verify its responsiveness and scalability under varying data loads empirically. For this purpose, we utilized three different open datasets to create our test base—DGT-Translation Memory [14], CCMatrix [3] and ParaCrawl [10]—selecting segments from the English–Spanish language pair. The English segments were targets for the fuzzy search queries, simulating realistic lookup operations in translation workflows. Four dataset sizes—100k, 500k, 1 million, and 10 million translation memory segments—were prepared to represent small, medium, large, and extra-large operational contexts. During a batch execution of 100 concurrent fuzzy search queries for each dataset, three key metrics were collected: the duration of fuzzy retrieval performed by Elasticsearch, the time taken for blockchain-based validation using Hyperledger Fabric, and the total end-to-end response time.

Table 1 shows descriptive statistics for these metrics across all test scenarios, including mean, median, standard deviation, and range. The results show that the system maintains consistent performance as data volume increases. Notably, even when processing a corpus of ten million segments, the system preserved real-time interaction standards: the median total query duration remained well below 250 ms, and the mean did not exceed 500 ms. These values are competitive, especially considering the significantly larger dataset. The results suggest that Elasticsearch indexing and the modular architecture continue to scale efficiently, and that Hyperledger Fabric maintained predictable performance behavior even under heavier loads.

Table 1. Summary of Evaluation Metrics by Dataset Size

Metric	Statistic	100k	500k	1M	10M
Elasticsearch Duration	Mean	189.30	360.87	218.21	237.57
	Median	63.00	99.00	63.00	173.00
	Min	9.00	27.00	11.00	6.00
	Max	664.00	1268.00	896.00	893.00
	Std	224.67	470.38	314.19	232.19
Hyperledger Duration	Mean	129.13	88.84	114.83	250.65
	Median	117.00	58.00	108.50	100.00
	Min	45.00	13.00	7.00	12.00
	Max	360.00	420.00	349.00	1190.00
	Std	53.17	71.28	55.95	337.73
Total Duration	Mean	318.43	449.71	333.04	488.22
	Median	187.50	161.00	176.50	234.00
	Min	73.00	58.00	40.00	80.00
	Max	969.00	1680.00	1167.00	1545.00
	Std	261.43	527.39	343.92	459.03

Figure 2 shows the statistical distribution of durations through box plots that display each metric across all four datasets. The distribution curves indicate that the system operates steadily because most queries remain concentrated within the lower quartiles. Even with 10M segments, the Hyperledger validation did not exhibit exponential growth, and the Elasticsearch retrieval remained tightly bounded, demostrating the robustness and predictability of the architecture.

The scatter plot in Fig. 3 shows total query durations against dataset size, with color-coding for each dataset. The figure confirms that most queries

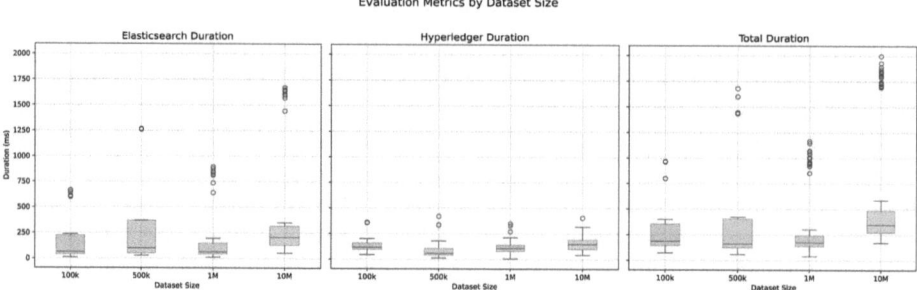

Fig. 2. Boxplots of Elasticsearch, Hyperledger, and Total Durations by Dataset Size.

remained under 500 ms even for the largest dataset. The 1 million segment dataset produced slightly better average response times than the 500k case, which can be attributed to improved Elasticsearch index optimization and effective caching behavior. Meanwhile, the results for 10 million segments were consistent and encouraging, with performance remaining well within practical thresholds for real-time systems.

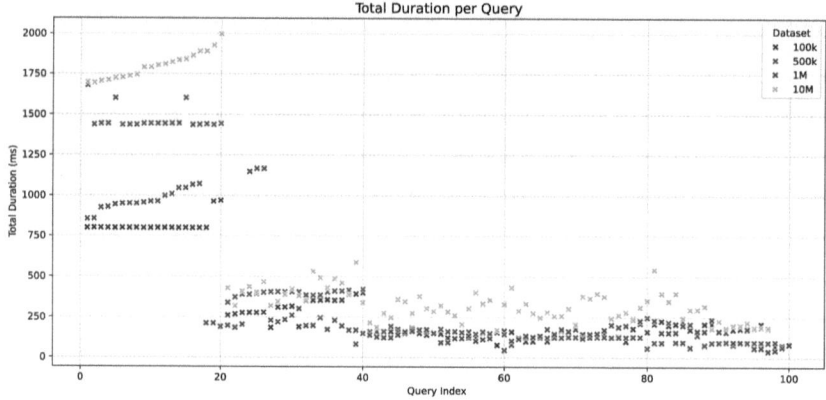

Fig. 3. Scatter plot of total query durations by dataset size (X-axis: queries 1–100).

The evaluation shows that the architecture meets its main performance goals, including low-latency fuzzy retrieval, predictable blockchain validation, and robust scalability. The results show that the system is ready for deployment in real-world high-demand translation memory environments where speed and data integrity are important.

8 Discussion

The evaluation results indicate that the proposed hybrid architecture effectively balances performance, scalability, and trust in the context of decentralized TM systems. The system consistently delivers sub-second response times, even as the dataset scales from 100k to 10M segments, which affirms its practical viability for real-time usage in professional translation environments.

A particularly noteworthy outcome is the minimal overhead introduced by the blockchain-based validation layer. Contrary to common assumptions that on-chain verification can severely impact performance, Hyperledger Fabric added only a moderate latency penalty—even for the 10M dataset, the median blockchain validation time remained around 100 ms. This validates the feasibility of integrating trust-enforcing mechanisms into high-throughput, user-facing applications without compromising responsiveness.

The architectural separation of concerns between retrieval and validation also proved critical. By allowing Elasticsearch to independently handle fuzzy matching while offloading integrity checks to the blockchain layer, the system benefits from both rapid search performance and tamper-evident validation. Moreover, the modular gateway-based approach facilitates seamless integration with existing CAT tools, enhancing the system's adaptability in enterprise translation workflows.

The proposed solution goes beyond technical performance to solve critical ethical and social issues. Conventional TM systems depend on centralized repositories, which restrict user autonomy while making data ownership and access control mechanisms unclear. By decentralizing the validation and metadata layer, the proposed system returns a degree of agency to the data contributors—translators, localization teams, and organizations—allowing them to retain control over their linguistic assets. This is particularly relevant in collaborative and cross-organizational environments where accountability, transparency, and respect for data sovereignty are critical.

The architecture provides privacy benefits when sensitive or proprietary content is involved. The system protects raw linguistic data by storing only hashed segment metadata on-chain, which enables audibility without revealing the original content and thus reduces the risk of data leakage or misuse. The system features match the increasing social demand for responsible data governance and digital trust.

From a research perspective, this work contributes a novel, implementable framework that bridges the gap between fuzzy search and blockchain-based trust mechanisms. It demonstrates how distributed systems and information retrieval techniques can be harmoniously combined to support performance and functionality, as well as the broader values of transparency, accountability, and ethical data use.

These contributions align with broader research priorities in database system design, privacy-aware architectures, and applied knowledge engineering. The system's scalability, extensibility, and socio-technical alignment reinforce its potential impact in academic research and practical, industry-driven applications.

9 Conclusion

This research introduced a hybrid architecture enabling efficient fuzzy search over TM segments within a decentralized and privacy-aware environment. By integrating Elasticsearch for high-performance approximate matching with Hyperledger Fabric for blockchain-based validation, the system addresses key challenges of scalability, data integrity, and trust in distributed linguistic data systems.

The system demonstrated strong performance characteristics through a comprehensive experimental evaluation on datasets ranging from 100k to 10M segments, maintaining sub-second query responses across all scales. The blockchain validation layer introduced only marginal overhead, validating its practicality for real-time applications without compromising responsiveness. These results affirm the proposed approach's technical soundness and readiness for deployment in production scenarios.

The research makes engineering contributions while simultaneously addressing ethical and social challenges that emerge from data governance in translation workflows. Traditional TM systems operate under platform provider control, reducing transparency and user autonomy. The proposed decentralized model grants translators and organizations cryptographic ownership of their contributions while smart contracts enable verifiable access histories. The new paradigm establishes fairer data ecosystems that support emerging digital sovereignty and accountability principles.

The architecture also supports privacy by design: only hashed metadata is stored on-chain, allowing for auditability without exposing sensitive linguistic content. Such design choices reflect a growing awareness of responsible data handling in collaborative environments and make the solution particularly suitable for cross-border or regulated translation contexts.

Several promising future research directions include integrating adaptive ranking algorithms using machine learning to improve match relevance, extending smart contract logic to enforce fine-grained access control, and deploying the system in real translation environments to evaluate usability and domain adaptation. Further exploration of decentralized identity frameworks could also enhance authentication while preserving privacy.

In conclusion, this research presents a scalable, interoperable, and ethically grounded architecture that bridges fuzzy search capabilities with decentralized trust. It demonstrates how modern database engineering can intersect with blockchain infrastructures to deliver performant systems that uphold transparency, privacy, and user empowerment—increasingly essential principles in the next generation of data-centric applications.

Declarations

Data Availability. The datasets analyzed during the current study are publicly available. Part of the data was obtained from the *DGT-Translation Memory* dataset, provided by the European Union, accessible at https://data.europa.eu/data/datasets/dgt-translation-memory?locale=en. Additional parallel corpora were extracted from the *OPUS* project, an open collection of translated texts maintained by the University of Helsinki, available at https://opus.nlpl.eu

Funding. This work was supported by the National Council for Scientific and Technological Development (CNPq), Brazil.

Competing Interest. The authors declare no competing interests.

References

1. Androulaki, E., et al.: Hyperledger fabric: a distributed operating system for permissioned blockchains. In: Proceedings of the Thirteenth EuroSys Conference, ACM (2018)
2. Azaria, A., Ekblaw, A., Vieira, T., Lippman, A.: MedRec: using blockchain for medical data access and permission management. In: Proceedings of the 2nd International Conference on Open and Big Data, IEEE (2016)
3. CCMatrix - Translation Memory Dataset. https://opus.nlpl.eu/CCMatrix/en&es/v1/CCMatrix. Accessed 30 June 2025
4. Hao, H., Huang, G., Liu, L., Zhang, Z., Shi, S., Wang, R.: Rethinking translation memory augmented neural machine translation. arXiv preprint arXiv:2306.06948 (2023)
5. Feng, T., Yang, P., Liu, C., Fang, J., Ma, R.: Blockchain data privacy protection and sharing scheme based on zero-knowledge proof. Wirel. Commun. Mob. Comput. **2022**(1), 1040662 (2022)
6. Hoang, C., Sachan, D., Mathur, P., Thompson, B., Federico, M.: Improving retrieval augmented neural machine translation by controlling source and fuzzy-match interactions. arXiv preprint arXiv:2210.05047 (2022)
7. Liu, H., Hunter, L., Kešelj, V., Verspoor, K.: Approximate subgraph matching-based literature mining for biomedical events and relations. PLoS ONE **8**(4), e60954 (2013)
8. McSherry, F.: An introduction to Elasticsearch. Springer, Information Retrieval Journal (2015)
9. Nguyen, T.L., et al.: Blockchain-empowered trustworthy data sharing: fundamentals, applications, and challenges. ACM Comput. Surv. **57**(8), 1–36 (2025)
10. ParaCrawl - Translation Memory Dataset. https://opus.nlpl.eu/ParaCrawl/en&es/v9/ParaCrawl. Accessed 30 June 2025
11. Peffers, K., Tuunanen, T., Rothenberger, M.A., Chatterjee, S.: A design science research methodology for information systems research. J. Manag. Inf. Syst. **24**(3), 45–77 (2007)
12. PoliLingua: Translation Memory Software and TM System. https://www.polilingua.com/blog/post/translation-memory-software-and-tm-system.htm. Accessed 11 May 2024
13. Reinke, U.: State of the art in translation memory technology. In: Rehm, G., Stein, D., Sasaki, F., Witt, A. (eds.) Language Technologies for a Multilingual Europe, pp. 55–84 (2018)
14. Steinberger, R., Eisele, A., Klocek, S., Pilos, S., Schlüter, P.: DGT-TM: A freely available translation memory in 22 languages. arXiv preprint arXiv:1309.5226 (2013)

15. Tezcan, A., Skidanova, A., Moerman, T.: Improving fuzzy match augmented neural machine translation in specialised domains through synthetic data. Prague Bull. Math. Linguist. **122**, 9–42 (2024)
16. Yujian, L., Bo, L.: A normalized Levenshtein distance metric. IEEE Trans. Pattern Anal. Mach. Intell. **29**(6), 1091–1095 (2007)
17. Zobel, J., Williams, H.E.: Search Engine Technology for Information Retrieval. ACM, ACM Computing Surveys (2006)
18. Zou, X., Zhan, Z., Xu, Z., Wu, Q., Cao, B.: Fuzzy keyword searchable encryption scheme based on blockchain. Information **13**(10), 517 (2022)

Query Answering and Education

Towards Sustainable DBMS: A Framework for Real-Time Energy Estimation and Query Categorization

Tidenek Fekadu Kore[1(✉)], David Sarramia[2], Myoung-Ah Kang[3], and François Pinet[1]

[1] Université Clermont Auvergne, INRAE, TSCF, Aubière, France
Tidenek_Fekadu.KORE@uca.fr, francois.pinet@inrae.fr
[2] Université Clermont Auvergne, CNRS/IN2P3, LPCA, Clermont-Ferrand, France
david.sarramia@uca.fr
[3] Université Clermont Auvergne, ISIMA, LIMOS, Aubière, France
kang@isima.fr

Abstract. Energy efficiency in database management systems (DBMS) is increasingly critical due to the rising computational demands of modern applications. Our work proposes a complete framework to analyze energy consumption. We developed a real-time monitoring framework that captures CPU and memory utilization during query execution and estimates energy consumption. We have implemented a query logging mechanism to track and analyze execution time. We propose an energy estimation model that computes power consumption using CPU utilization metrics and query categorization based on energy usage profiles. We studied the correlation between execution time and energy consumption using Pearson correlation. We propose a power-based classification of SQL query types, enabling more energy-aware optimization strategies. The result of our analysis highlights the opportunities for power-aware query optimization, making DBMS operations green computing and efficient.

Keywords: Energy-Efficient Computing · Database Management Systems (DBMS) · Query Optimization · Power Consumption Analysis · Energy Estimation Model · Green Computing · Workload Profiling

1 Introduction

Data-driven applications rapidly grow and demand an efficient database system to handle large volumes of queries. Database systems are integrated into software architecture. Accordingly, optimizing their performance is essential for performance and energy efficiency, but traditionally, the performance of database queries is measured by execution time without considering energy efficiency, as a growing importance in sustainable computing environments [1, 2]. Usually, power consumption is analyzed at the system level through TPC-C benchmarks developed by the Transaction Processing Performance Council to evaluate the performance of online transaction processing (OLTP) systems [3].

© The Author(s), under exclusive license to Springer Nature Switzerland AG 2026
G. Bergami et al. (Eds.): IDEAS 2025, LNCS 15928, pp. 171–183, 2026.
https://doi.org/10.1007/978-3-032-06744-9_13

This work emphasizes the energy and costs associated with large-scale database operations and primarily focuses on the system-level energy efficiency, leaving the query-level energy consumption unclear [3]. Fine-grained energy profiling was introduced for developing power-aware applications, in which energy consumption is monitored at a detailed level to optimize overall system performance [4].

Data centers are expected to improve energy efficiency by reducing energy cost, carbon footprint, database efficiency, operational costs, and environmental sustainability [5]. The challenge lies in the accuracy of capturing real-time performance metrics and relating them to energy consumption. It needs precise monitoring of CPU and memory usage, logging of query execution time, and also a model to estimate energy consumption on the hardware performance statistics [6].

Database performance analysis predominantly focuses on query execution time, indexing strategies, and algorithmic optimizations [7]. However, energy consumption metrics are rarely incorporated into these performance evaluation frameworks. Traditional monitoring tools either lack the holistic approach needed to capture detailed usage statistics or are not designed to correlate performance metrics with energy consumption [8]. The available tools make obtaining a holistic view of performance and energy efficiency challenging.

In this paper, we propose a comprehensive methodology to simultaneously analyze both performance and energy efficiency in database systems. While there are existing solutions focused on query optimization to reduce execution time [9], few consider the energy consumption associated with query processing [10]. This is a significant gap because it results in enormous energy consumption, including CPU and memory usage during query execution. The cost encompasses IT equipment (i.e., CPU, memory, disk, etc.), cooling infrastructure, electricity, lighting, and environmental implications [11].

The contributions proposed in this paper are the following. Our study proposes a generalizable methodology to analyze the efficiency of SQL queries in terms of both energy consumption and execution time, aiming to support energy-aware database optimization strategies that can be applied across different systems and instances. This methodology is demonstrated using a specific dataset, showcasing its practicality and effectiveness.

We study the relationship between query execution time and energy consumption through correlation analysis. To categorize queries based on their power consumption, we employ a power-based categorization approach that groups queries according to their power usage profiles. Additionally, we developed a real-time monitoring system to capture CPU and memory usage statistics during database query execution. A logging mechanism to track query start and end times, allowing for execution time tracking, result visualization, and export. Furthermore, energy consumption estimation based on CPU utilization metrics provides valuable insights into power efficiency, helping optimize resource usage in database systems.

These contributions provide valuable insights into both performance and energy efficiency, supporting more sustainable and cost-effective database management practices [10]. The rest of this study is organized as follows. In the next section, we present related works. In Sect. 3, we present the technical details of our energy cost model. In Sect. 4, we present the experimental results of our models. Finally, we present the conclusions and future work.

2 Related Work

In this section, we present the main concepts related to our topic in the fields of query performance optimization, energy-efficient computing, performance and energy monitoring tools, and correlation between query execution time and energy consumption.

2.1 Query Performance Optimization

Query optimization in databases primarily focused on reducing query execution time. Methods like indexing (i.e., B-trees, hash indexes, and bitmap indexes) enhance data retrieval efficiency, while cost-based and heuristic-based optimizers evaluate execution plans to minimize computational costs. Caching mechanisms or materialized views reduce redundant computations and enhance response times [12]. However, traditional approaches focused on improving execution time and not on the associated energy consumption, which is important as database scale [7].

An online scheme dynamically adjusts model parameters based on statistical signal modeling results to optimize query plans for energy efficiency. It also analyzes various relational operations, such as JOIN and SELECT, to evaluate their impact on energy consumption across different execution times [13].

2.2 Energy-Efficient Computing

Energy-efficient computing is considered in large-scale data centers and cloud environments, with the growing concern about operational costs and environmental impact. Dynamic Voltage and Frequency Scaling (DVFS) dynamically tunes CPU power consumption based on workload demands [1]. Energy-aware task scheduling optimizes workload distribution to reduce power consumption [9, 13]. Some studies incorporated energy into query engines by modifying query execution plans [9], often requiring changes to database internals. Our approach proposes a monitoring framework that captures energy usage without altering database internals, enabling large-scale reuse.

2.3 Performance and Energy Monitoring Tools

Tools such as MySQL's Performance Schema and PostgreSQL's pg_stat_statements and EXPLAIN ANALYZE are commonly used to collect execution statistics, but they fail to provide detailed energy metrics [15]. Meanwhile, energy profiling tools such as Intel's Running Average Power Limit (RAPL) and PowerAPI measure CPU power consumption but do not link these measurements to specific database query executions [8, 17].

DBJoules tools measure the energy consumption of create, read, update, and delete (CRUD) operations across different databases but do not analyze the correlation with execution time, providing insights into how to measure energy consumption at the query level [15]. Because the integrated tools that combine performance and energy monitoring are a critical gap, we propose a comprehensive solution for monitoring both performance and energy efficiency simultaneously.

2.4 Correlation Between Query Execution Time and Energy Consumption

Despite existing work on energy efficiency in computing, few studies have explored the relationship between query execution time and energy consumption at a granular level. Research in different environments, such as MySQL, PostgreSQL, MongoDB, and Couchbase, has primarily focused on energy analysis on individual database queries but lacks a detailed quantitative correlation analysis [18].

There is a general cloud-based approach to comparing energy usage, but it does not specifically the relationship between individual query execution characteristics and energy consumption. This makes it unclear, as it fails to break down query-level factors that impact energy consumption and lacks time-based correlations [17]. We propose to perform an analysis of the correlation between CPU energy consumption and query execution time. Additionally, we propose an approach for grouping queries based on their energy consumption, providing insights into query-level power efficiency. The goal is to show the behavior of different query families in terms of energy consumption.

3 Methodology

This section describes the methodology we propose: the system design, implementation, and experimental setup for analyzing database query performance and energy efficiency.

3.1 System Design and Architecture

Our system architecture is designed to provide real-time performance and energy efficiency analysis of database query executions. The architecture includes five interconnected modules that capture performance metrics, estimate energy consumption, and visualize the collected data.

To implement the proposed monitoring and analysis framework, we utilized a combination of Python-based tools and system-level APIs. The psutil library was employed for real-time monitoring of CPU and memory utilization [21, 22]. For interaction with the PostgreSQL database, we used the psycopg2 API, enabling query execution and logging within the system [14]. The energy estimation model used CPU usage data to estimate power consumption. Visualization of results was handled using matplotlib and seaborn for plotting performance metrics, while pandas was used for structured data handling and analysis.

As shown in Fig. 1, the architecture includes the following key components: The Real-Time Monitoring Module, Query Logging Mechanism, Energy Estimation Model, and Visualization Module.

Figure 1 describes the architecture of our system, with a real-time monitoring module that tracks CPU and memory utilization. The data from this module is used by an energy estimation model to calculate energy consumption based on CPU utilization percentages. Additionally, a database system logs queries with start and end times using treads and then visualizes them along with the energy estimation results.

We present a comprehensive methodology to evaluate the energy consumption and performance of database queries. Queries are categorized (e.g., selection, join, aggregation) for structured analysis and executed with PostgreSQL's EXPLAIN ANALYZE

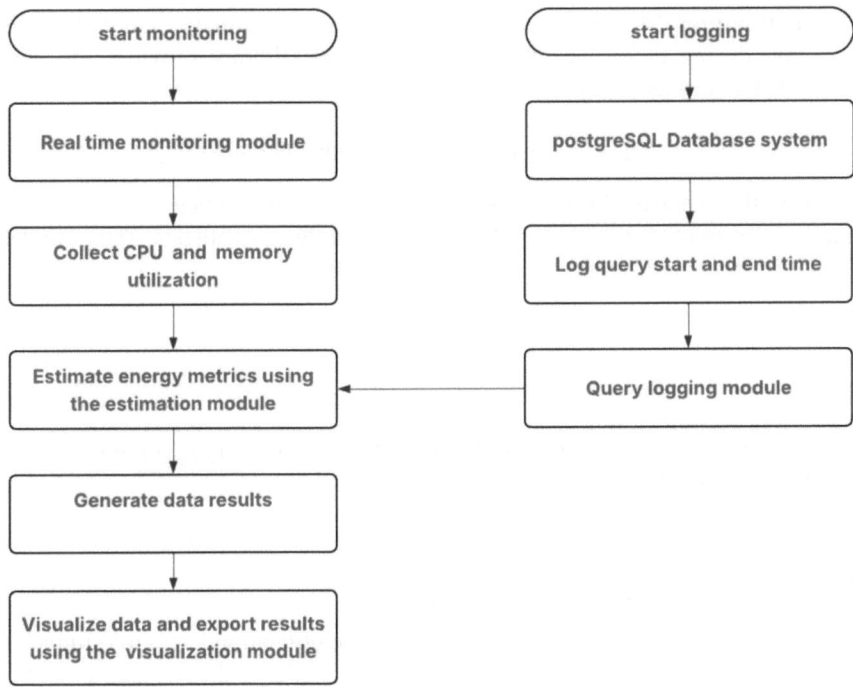

Fig. 1. System architecture for real-time query performance and energy consumption monitoring.

to obtain detailed execution plans. Real-time monitoring and logging modules capture system resource usage and execution timing, which are integrated to estimate per-query energy consumption. Finally, collected data undergoes processing, analysis, and visualization to compare performance metrics.

The Real-Time Monitoring Module. The goal is to capture real-time CPU and memory utilization by recording CPU and memory usage statistics between the start to end times, and return the average of the usage during the execution of each query using the process utilities (psutil) library.

The Real-Time Monitoring Module continuously tracks CPU and memory utilization during database query execution. This module ensures that performance data is collected with minimal overhead (i.e. 0.1 s).

Query Logging Mechanism. In our approach, each query execution is logged with start and end timestamps to calculate execution times.

$$\text{Execution time(s)} = \text{query end time} - \text{query start time} \quad (1)$$

where execution time is the duration to execute each query.

The Query Logging Mechanism logs query execution metrics such as CPU usage, memory usage, query start time, query end time, execution time, energy consumed by CPU, and energy consumed by memory. The captured execution data supports analysis of how query structure affects performance and energy consumption.

Energy Estimation Model. The CPU energy model estimates the CPU energy consumption during query execution [20]. The psutil library provides the current system-wide CPU usage as a percentage [21, 22].

$$\text{CPU usage (\%)} = \text{psutil.cpu\_percent(interval} = 0.1) \tag{2}$$

It returns the average CPU usage across all cores as a percentage.

Energy consumption is estimated based on system usage statistics. Thus, we define that the CPU energy consumption is computed as follows

$$ECPU = TDP * WCPU * t \tag{3}$$

Where:

- ECPU: the total energy consumption of the CPU during query execution
- WCPU: the CPU utilization percentage
- t: the CPU loading time
- TDP (thermal design power) is the average power a CPU is expected to consume under normal maximum load.

This model uses TDP specifications and real-time utilization levels to estimate total CPU energy consumption during query processing. TDP is the average power consumption specific to a CPU model [22]. For example, for the 13th Gen Intel® Core™ i7-13700H, which has a TDP of 45W, the CPU power consumption is estimated based on processor power specifications and real-time utilization levels [14]. The total energy consumption is then calculated using the CPU utilization percentage and execution time. Memory energy model estimates the energy consumption by retrieving the current system-wide memory usage using the psutil library. The psutil.virtual_memory() function provides information about the system's total, used, and free memory, which is needed for estimating power consumption during query execution [21]. To obtain the total amount of used physical memory, we extract the used memory value in bytes and convert it into megabytes (MB). RAM power consumption is considered proportional to the amount of allocated power by the current running process. The RAM energy consumption is determined using the equation proposed by [22]. The memory usage was measured using the psutil library in Python:

$$\text{memory usage (MB)} = (\text{psutil.virtual\_memory().used})/1024 * 1024 \tag{4}$$

The energy consumption of RAM (ERAM) is then estimated using:

$$ERAM = 0.375 * MRAM * t \tag{5}$$

Where:

- ERAM: Estimated energy consumption of RAM (in Joules),
- MRAM: Allocated memory (in MB), obtained using psutil,
- *t*: Data loading time (in seconds),

- The constant 0.375 (in W/GB) represents the estimated specific energy consumption of DDR3 and DDR4 memory modules [20–22].

Memory power consumption is proportional to the amount of power allocated by the current running processes, where higher memory usage typically leads to increased energy consumption [22].

Data Visualization Module. The collected performance metrics and energy consumption data are analyzed using visualization Python libraries such as Matplotlib, Pandas, and NumPy. Generates graphs to show the relation between CPU and memory energy consumption over query execution time across different query types. The Visualization Module generates visual representations of performance and energy consumption data.

It enables database administrators and researchers to identify patterns, optimize queries, and reduce energy consumption.

3.2 Development Stack and Tools

In this section, we present the platform, experimental setup, database management framework, programming language, and profiling tools.

CEBA Platform. Our methodology was experimented on data of a real cloud platform called the Environmental Cloud for the Benefit of Auvergne (CEBA) [23]. This platform stores data produced by sensor networks measurements. CEBA is a cloud-based infrastructure designed to support agricultural activities by providing environmental data and analytical tools to enhance agricultural sustainability by leveraging environmental data such as weather conditions (i.e. temperature, humidity, precipitation), soil moisture, geospatial data, and other relevant parameters. The platform includes cloud storage, database systems, and data lakes to store diverse data types, ensuring scalability and accessibility for various users and applications. CEBA offers a rich dataset for researchers to develop new agricultural technologies and practices. It stores data received from a network of sensors, specifically designed to handle time-series data, which are georeferenced data. The database size used stands at approximately between 2.5 GB - 10GB on the CEBA platform. A total of 2.5 GB of data was used for this analysis. This setup allows us to conduct all experiments on a single database server, ensuring efficient processing and management of our tasks.

The experimented queries are mainly based on two tables: connecsens.json_montoldre_row (2,235,825 rows) and connecsens.json_file (7,474 rows). The json_montoldre_row table contains JSON data that includes several main attributes.

Hardware Specification. Our experimentation environment runs on a setup of a 13th Gen Intel(R) Core (TM) i7-13700H processor clocked at 2.40 GHz and 16.0 GB of RAM (i.e., 14 cores, 20 threads). Operating on Windows 11 Professional 64-bit.

Software Stack. PostgreSQL can manage JSON data. Within CEBA, the sensor database is organized around its associated sensor networks, employing various schemas tailored to the received data structure. Presently, there are two schemas implemented:

one for delimited files and another for the Connecsens network utilizing JSON files. JSON has been adopted as the preferred format for managing IoT data on the CEBA platform, leading to its extensive usage within their database. Occasionally, there is a need to convert raw data files into JSON format. In PostgreSQL, both JSON and JSONB (JavaScript Object Notation Binary) are available for storing raw JSON data. JSONB was chosen due to its broader array of query and comparison capabilities compared to JSON.

Python (v3.11.9) is the primary programming language in our work due to its diverse libraries (i.e., psutil v6.1.0, matplotlib v3.9.2, numpy v2.1.2) and frameworks that facilitate integration with PostgreSQL (v17.4), storing results in CSV files, plotting graphs for the results, and running workloads on PostgreSQL.

The system was integrated with PostgreSQL using Python, psycopg2 (v2.9.10) library to track query execution while minimizing overhead. Query execution metadata was captured through EXPLAIN ANALYZE statements.

3.3 Implementation

We started the implementation of a real-time monitoring algorithm by using the proposed System architecture discussed in Sect. 3.1 to collect and analyze key performance metrics such as query execution time, CPU usage, memory usage, and Estimated Energy Consumption. We tested and validated the algorithm using different workloads. Here are examples of queries that can be executed: Aggregate functions, Arithmetic query, and Join.

We executed each query to avoid the impact of the previous query cache. During the execution of the query, CPU and memory utilization are logged in real-time, logging execution time and estimated energy consumption. Pearson correlation analysis is used to evaluate the correlation between query execution time and energy consumption. This statistical method quantifies the linear relationship between them to determine whether execution time is a predictor of energy consumption [24]. The system is designed to correlate query execution time with energy consumption, estimate power efficiency based on CPU utilization metrics, and categorize queries by energy usage profiles.

4 Result and Discussion

4.1 Performance Metrics

This section evaluates our approach using read-only workloads, analyzing CPU energy consumption (ECC in J), execution time (ET in s), total accessed rows (TAR), and output rows (OR). While our energy model supports memory metrics, they are excluded here due to minimal variation and limited impact on comparative analysis.

Basic queries such as SELECT * FROM exhibit lower ET but moderate ECC. For instance, SELECT (*) FROM data-node-timestampUTC executes in 38.56 s with 105.29 J. Conversely, complex queries like SELECT DISTINCT FROM or those with WHERE clauses generally increase ECC due to added computational overhead.

A trade-off is evident: reduced ET does not always equate to lower ECC. Some fast queries involve intense CPU activity (e.g., indexed lookups), leading to higher energy use. For example, GROUP BY queries tend to minimize both ET and ECC (e.g., 1.58 s and 2.64 J), while non-aggregated SELECT FROM versions can be significantly more demanding.

Aggregation functions (AVG, SUM) introduce further overhead. On data-node-batteryVoltage, SELECT AVG FROM GROUP BY yields ET = 8.09 s, ECC = 32.31 J, whereas the plain SELECT requires only 0.88 s and 2.38 J. Similarly, MIN and MAX are efficient alone but costlier when combined with GROUP BY (e.g., ECC jumps from 1.15 J to 25.09 J for data-temperature).

Casting operations (CAST) modestly increase ECC and ET, particularly when paired with grouping. JOIN operations are especially expensive unless constrained with LIMIT. For instance, JOIN on json_montoldre_row.data without LIMIT consumes 109.37 J (ET = 38.70 s), while the limited version uses only 28.91 J (ET = 5.10 s).

Arithmetic operations also vary in cost: multiplication and division have higher ECC and ET compared to addition or subtraction. For example, on applicationID, multiplication results in ECC = 4.44 J, and division leads to the highest ET (1.72 s).

In summary, query complexity significantly influences both execution time and energy use. Optimization efforts should prioritize minimizing grouping and computational operations where feasible, particularly in energy-constrained systems.

4.2 Analyzing SQL Query Efficiency: TAR vs Output Rows vs Power Usage

The 3D relationship between input rows, output rows, and power consumption (ECC/ET) captures how different SQL query structures impact energy efficiency and performance. The results show that queries with similar output cardinalities can have different power profiles depending on their query design and complexity. For instance, SELECT * FROM queries consume more energy and time than their SELECT FROM counterparts, even when returning the same output sizes. This indicates that query formulation (i.e., selecting all versus specific columns) plays a more significant role in power consumption than the output volume. Conversely, some queries (such as SELECT FROM WHERE) have a balance by reducing execution time and energy because of filtering. This shows that query design, not just data volume, is a key to energy efficiency.

The energy consumption analysis reveals distinct patterns tied to query complexity. Figure 2 presents a 3D plot of power consumption, showing that SELECT and WHERE queries are energy-efficient, while SELECT GROUP BY and SELECT DISTINCT incur higher power usage due to greater computational demands. Low-power queries form tight clusters, indicating efficiency, whereas high-power queries are more dispersed, reflecting their resource-intensive nature.

Figure 3 illustrates power consumption for aggregate queries (AVG, SUM, MIN, MAX), with and without the GROUP BY clause. Aggregations without grouping consume significantly less power, highlighting the overhead introduced by grouping even when accessing a similar number of rows. This suggests that avoiding GROUP BY in large datasets can improve energy efficiency.

Figure 4 analyzes power usage in JOIN, CAST, and arithmetic queries. Results show that joins, especially when combined with casting or GROUP BY, lead to the highest

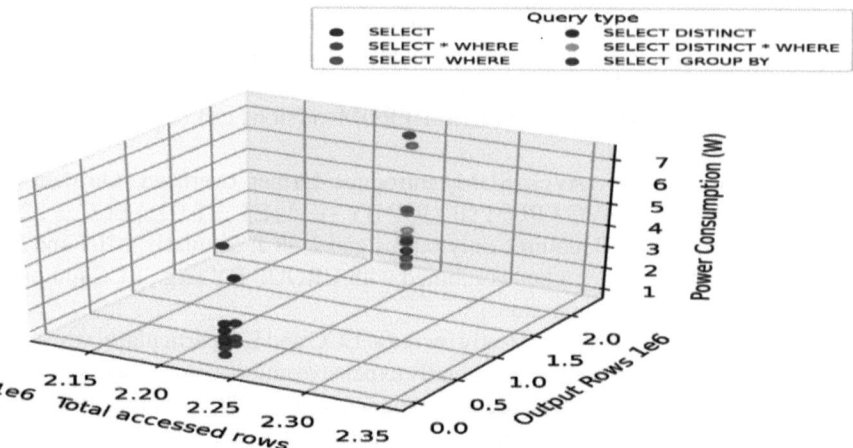

Fig. 2. 3D visualization of SQL query types based on TAR, output rows, and power consumption.

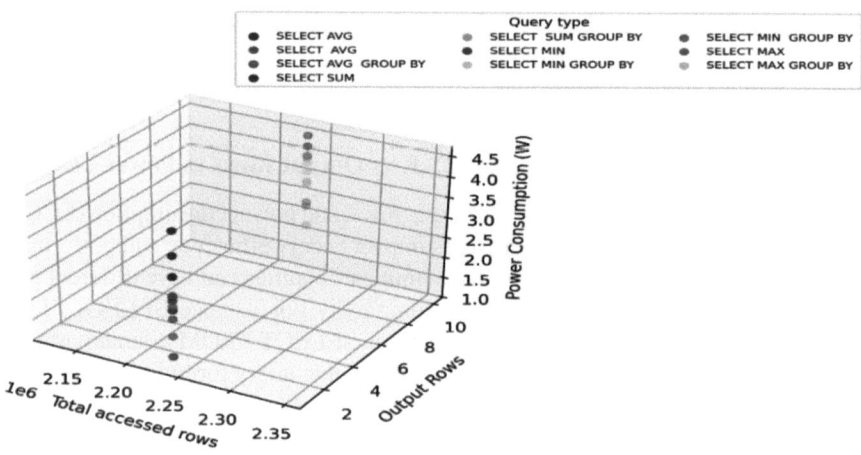

Fig. 3. 3D visualization of SQL query types based on TAR, output rows, and power consumption.

power consumption due to increased computational complexity. Queries with more output rows also show higher energy usage, underlining the importance of optimizing joins and minimizing unnecessary data transformations.

The 3D plots effectively visualize these patterns, helping to identify which query types offer optimal trade-offs between power and performance, thereby guiding energy-efficient query design strategies.

4.3 Power-Based Categorization of SQL Query Types

To analyze the energy efficiency of different query types, we categorized them based on their power consumption per execution time (ECC/ET) ratio, which reflects the average power usage (in watts) during query execution. Grouping queries with similar ECC/ET

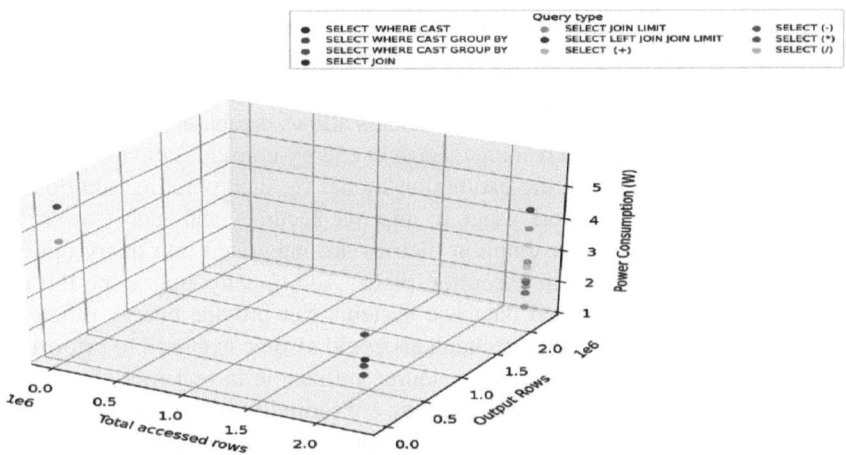

Fig. 4. 3D visualization of SQL query types based on TAR, output rows, and power consumption.

values helps identify patterns and supports energy-aware optimization strategies. Based on the observed ratios, we defined three categories:

- Low power consumption (ECC/ET < 2): e.g., SELECT, GROUP BY
- Medium power consumption ($2 \leq$ ECC/ET < 5): e.g., SELECT *, DISTINCT, DISTINCT *, CAST, arithmetic, aggregation
- High power consumption (ECC/ET $\geq$ 5): e.g., JOIN, LEFT JOIN

These findings show that more complex queries, such as JOINs, tend to consume more power per second of execution. Categorizing queries by ECC/ET facilitates the development of energy-efficient execution plans, especially in large-scale or resource-constrained environments.

5 Conclusion and Future Work

We presented our work on creating a real-time monitoring and energy estimation framework for evaluating query execution performance and power consumption in DBMS. The experiment involved 2,243,299 rows and 9 attributes. By tracking CPU and memory usage, logging execution times, and estimating energy consumption, we associate correlations between query execution time and energy consumption. Our findings showed that query complexity impacts energy efficiency, and queries can be categorized for optimized scheduling and resource management. The proposed energy estimation approach provides a practical method to estimate energy usage besides hardware power meters. This study contributes to energy-efficient query execution strategies, contributing to sustainable computing practices in database systems. Additionally, it is possible to use this framework for non-relational databases.

Analyzing an average energy coefficient for each query family enables us to understand the energy behavior associated with different types of database operations. This coefficient characterizes the energy efficiency of query execution time, provides insights for energy-aware query optimization, and workload scheduling. Queries within the same family often have similar energy-to-time ratios, indicating a predictable energy profile. However, variations may occur due to differences in query structure, input size, or execution plans. Overall, having such coefficients allows database systems to optimize performance and cost-effective resource usage in energy-conscious environments.

Future work can focus on the estimation model by incorporating additional factors such as disk I/O and network usage to improve accuracy and energy-aware query optimization, where power consumption metrics are integrated into query optimizers to suggest more energy-efficient execution plans. Additionally, upgrading the framework for distributed cloud-based database systems can provide energy efficiency at scale. Dynamic query scheduling mechanisms could also be explored to prioritize low energy-consuming queries, optimizing resource utilization in real time. Furthermore, machine learning techniques to predict energy consumption based on query structure, workload patterns, and historical data could improve the adaptability of power-aware DBMS solutions. It will contribute to developing more energy-efficient and sustainable DBMSs.

Acknowledgements. This research was financed by the French government IDEX-ISITE initiative 16-IDEX-0001 (CAP 20-25).

References

1. Guo, B., Yu, G., Yang, D., Leng, H., Liao, B.: Energy-efficient database systems. A systematic survey. ACM Comput. Surv. **55**(6), 1–38 (2023)
2. Xu, Z., Tu, Y.-C., Wang, X..: Exploring power-performance tradeoffs in database systems. In: Proceedings for the 26th IEEE International Conference on Data Engineering. (ICDE) (2010)
3. Poess, M., Nambiar, R.O.: Energy cost, the key challenge of today's data centers: a power consumption analysis of TPC-C results. In: Proceedings of the VLDB Endow, pp.1229–1240 (2008)
4. Kansal, A., Zhao, F.: Fine-grained energy profiling for power-aware application design. ACM Sigmetrics Perform. Eval. Rev. **36**, 26–31 (2008)
5. Da Costa, G., Pierson, J.-M., Hérault, H., Oleksiak, A., Piotr, W.: PowerAPI: a software library to monitor the energy consumed at the process level. ERCIM News (2013)
6. Garcia-Molina, H., Ullman, J.D., Widom, J.: Database systems: the complete book, 2nd edn. Prentice Hall, Upper Saddle River, NJ, USA (2008)
7. MySQL Performance Schema. https://dev.mysql.com/doc/refman/8.0/en/performance-schema.html. Accessed 25 Dec 2024
8. Guttman, A.: R-trees: a dynamic index structure for spatial searching. In: Proceeding of the ACM SIGMOD International Conference Management of Data, pp. 47–57. ACM, Boston, MA, USA (1984)
9. Chen, Y., Shen, B., Zhang, Y., Jin, X., Yu, Y.: A green framework for DBMS based on energy-aware query optimization and energy-efficient query processing. J. Netw. Comput. Appl. **84**, 118–130 (2017)
10. Behzadnia, P., Tu, Y.-C., Zeng, B., Yuan, W.: Energy-aware disk storage management: online approach with application in DBMS. arXiv preprint arXiv:1703.02591, (2017)

11. Chaudhuri, S.: An overview of query optimization in relational systems. In: Proceeding of the ACM Symposium Principles of Database Systems (PODS), pp. 34–43. Seattle, WA, USA (1998)
12. Xu, Z., Tu, Y.-C., Wang, X.: Online energy estimation of relational operations in database systems. IEEE Trans. Comput. **64**(11), 3223–3236 (2015)
13. Lang, W., Patel, J.M.: Towards eco-friendly database management systems. In: Proceedings of the Conference Innovative Data Systems Research (CIDR), pp. 1–8, Asilomar, CA, USA (2009)
14. Intel Corporation: Intel® 64 and IA-32 Architectures Software Developer's Manual. https://www.intel.com/content/www/us/en/developer/articles/technical/intel-sdm.html. Accessed 25 Jan 2025
15. Lella, H.S., Manasa, K., Chattaraj, R., Chimalakonda, S.: DBJoules: an energy measurement tool for database management systems. arXiv preprint arXiv:2311.08961 (2023)
16. Lella, H.S., Manasa, K., Chattaraj, R., Chimalakonda, S.: Towards comprehending energy consumption of database management systems: a tool and empirical study. In: Proceedings of the 28th International Conference Evaluation and Assessment in Software Engineering (EASE) (2024)
17. Bani, B., Khomh, F., Guéhéneuc, Y.-G.: A study of the energy consumption of databases and cloud patterns. In: Proceedings of the International Conference Service-Oriented Computing (ICSOC), pp. 606–614. Springer Nature, Banff, AB, Canada (2016)
18. Liu, X., Wang, J., Wang, H., Gao, H.: Generating Power-Efficient Query Execution Plan. In: Proceedings of 2nd International Conference on Advances in Computer Science and Engineering (CSE). Springer Nature, Los Angeles, CA, USA (2013)
19. Budennyy, S., et al.: Eco2AI: carbon emissions tracking of machine learning models as the first step towards sustainable AI. arXiv preprint arXiv:2208.00406 (2022)
20. Rodriguez-Martinez, M., Valdivia, H., Seguel, J., Greer, M.: Estimating power/energy consumption in database servers. J. Parallel Distrib. Comput. **6**, 112–117 (2011)
21. Maevsky, D.A., Maevskaya, E.J., Stetsuyk, E.D.: Evaluating the RAM energy consumption at the stage of software development. In: Green IT Engineering: Concepts, Models, Complex Systems Architectures, pp. 101–121. Springer Nature (2016)
22. Henderson, P., Hu, J., Romoff, J., Brunskill, E., Jurafsky, D., Pineau, J.: Towards the systematic reporting of the energy and carbon footprints of machine learning. J. Mach. Learn. Res. **21**, 1–43 (2020)
23. Sarramia, D., Claude, A., Ogereau, F., Mezhoud, J., Mailhot, G.: CEBA: A data lake for data sharing and environmental monitoring. Sensors **22**(7), 2733 (2022)
24. Rodgers, J.L., Nicewander, W.A.: Thirteen ways to look at the correlation coefficient. Am. Stat. **42**(1), 59–66 (1988)

Context-Aware Visualization for Explainable AI Recommendations in Social Media: A Vision for User-Aligned Explanations

Banan Mohammad Alkhateeb[1,2](✉) and Ellis Solaiman[1]

[1] Newcastle University, Newcastle Upon Tyne NE1 7RU, UK
{b.m.a.al-khateeb3,ellis.solaiman}@newcastle.ac.uk
[2] King Faisal University, Alahsa 31982, Kingdom of Saudi Arabia

Abstract. Social media platforms today strive to improve user experience through AI recommendations, yet the value of such recommendations vanishes as users do not understand the reasons behind them. This issue arises because explainability in social media is general and lacks alignment with user-specific needs. In this vision paper, we outline a user-segmented and context-aware explanation layer by proposing a visual explanation system with diverse explanation methods. The proposed system is framed by the variety of user needs and contexts, showing explanations in different visualized forms, including a technically detailed version for AI experts and a simplified one for lay users. Our framework is the first to jointly adapt explanation style (visual vs. numeric) and granularity (expert vs. lay) inside a single pipeline. A public pilot with 30 X users will validate its impact on decision-making and trust.

Keywords: Explainable AI · AI Trust · Social Media

1 Introduction

AI-based recommendation systems now drive user experience across all major social media platforms. Facebook, for example, uses various AI tools for content generation, friend suggestion, and ad personalization. Similarly, Instagram and X (Twitter) use AI for content generation and ad personalization. LinkedIn likewise suggests job-related content to its users [9, 19]. Yet users rarely understand why certain content is shown to them, leading to concerns and doubts about AI usage and a lack of trust in AI recommendations. Figure 1 shows a simple contrasting scenario of a black-box recommendation compared to a transparent recommendation. It illustrates how the user perceives the same recommendation in two different situations.

Although social media platforms have tried to incorporate explainability, current approaches ignore user diversity by providing uniform explanations for all. Additionally, their explanations are not tailored to user categories [1, 9, 17]. Therefore, personalized explanations based on user types need to be explored to build appropriate trust among all users and stakeholders. Also, the scarcity of XAI research in the context of social media

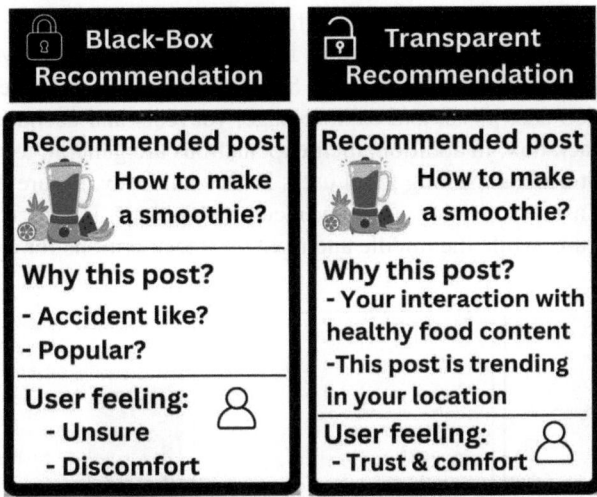

Fig. 1. Black-box recommendation vs. transparent recommendation

is another motive to do this work. This paper presents a vision for a visual explanation system tailored to diverse user needs, bridging the gap between AI decisions and human trust in social media. The key contributions of this vision paper are: **1)** A problem framing that highlights the limitations of one-size-fits-all explanations for diverse social media users. **2)** A novel phased framework for generating context-aware, visually personalized explanations tailored by user expertise and situational context. **3)** A planned evaluation strategy using trust, decision-time, and usability metrics, to validate the framework in a public pilot study on X (Twitter).

The rest of the paper is structured as follows: we present background and related work in Sect. 2, our vision and proposed framework in Sect. 3, our case example and data strategy in Sect. 4, key challenges and future direction in Sect. 5, and Sect. 6 is a conclusion of the paper.

2 Background and Related Work

2.1 Current State of XAI in Social Media

Popular social media platforms, such as Facebook, Instagram, and X (Twitter), use different technologies to build and improve their recommendation systems. While the exact algorithmic details are not explicitly shared with the public, these platforms are trying to be keen to consider transparency, user control, and explainability. Exploring the three mentioned platforms reveals that they give users some control over influencing and customizing their AI recommendations. For example, X users can manage their ad appearance and click on "not interested in this" to hide it. Likewise, Instagram users can adjust their ad and content preferences. Similar features can also be found on Facebook.

When it comes to explainability, the three mentioned platforms provide users with some explanation options, such as "why you're seeing this ad". Nevertheless, their explanations are insufficient and are not offered for all types of AI recommendations. For instance, X only provides explanations for ad recommendations, while it does not explain post and account recommendations. Similarly, Instagram offers explanations for their recommended ads, posts, and explore page, yet reels and account recommendations remain a black box. In addition, their explanations are general, textual, static, and communicate just basic reasoning to all users in the same way. Figure 2 shows sample screens of X and Instagram options for some recommendations. The three screens contain a controlling feature, while the middle and right ones lack options for explainability.

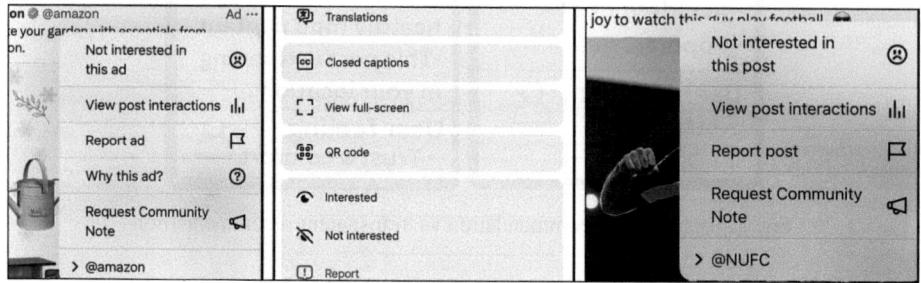

Fig. 2. Existing user control & explanation gaps on X and Instagram

Many researchers are dedicated to addressing social media explainability issues in their studies. One example is the Recommendation and Interest Modeling Application (RIMA) [7] that recommends Twitter content to users, showing on-demand explanations in three levels. However, RIMA does not emphasize personalization by user type or context, which is the focus of the proposed system. A summary of the major explainability features on the mentioned social media platforms is presented in Table 1.

Table 1. Explainability features in social media platforms vs. the proposed system

Platform	On-demand	Graph-based /visualized	Textual	Configurable	Personalized by user type
Facebook	Yes	No	Yes	No	No
Instagram	Yes	No	Yes	No	No
Twitter	Yes	No	Yes	No	No
RIMA	Yes	Yes	Yes	Yes	No
Proposed Sys.	Yes	Yes	Yes	Yes	Yes

2.2 Strengths and Limitations of XAI Methods

Model-agnostic explainable AI methods are widely used today to explain AI predictions. They are post-hoc techniques and are flexible to be used with any AI model. They are also capable of providing different forms and representations of an explanation [13].

Examples of such methods include LIME, SHAP, and Counterfactual Explanations. While these methods can contribute to enhancing the transparency of a black-box model, each one has its strengths and limitations.

SHAP is widely used in practice, and it excels at its global and local abilities. It can explain individual predictions as well as the general behavior of a model. The consistency and efficiency of its explanations make it suitable for even more complex models. Its explanations are also supported with visuals such as summary, dependence, and force plots [14, 16]. However, its high computational cost and complexity are significant drawbacks, limiting its applicability when targeting non-technical users [15].

LIME explanations are local and focused on interpreting individual AI decisions. This approach gives a quick insight into a single model prediction, explaining the influence of the model's features. LIME's explanations are usually supported with raw visuals, such as bar plots. This method is distinguished by its simplicity, ease of implementation and integration, as well as flexibility, making it very effective in many applications. Its output can be customized to suit various user needs, so generating tailored explanations that are user-friendly is possible. For instance, raw output of feature weights, which is too complex for non-technical users, can be transformed into a simple plain language with the use of icons and colors. Nevertheless, LIME may be sensitive to the variation of data samples, which may cause inconsistency or randomness of explanations. It is also limited in providing interactive visualizations [4, 10, 12, 15].

Contrastively, the counterfactual explanation method offers intuitive explanations by informing the user about the small input changes that can be made to generate different predictions, rather than explaining how the model works. Its explanations are basically what-if scenarios (if this input is different, the model output would change), which make them more useful in perceiving the reasoning behind an AI decision [21, 22]. However, many researchers criticize its applicability in practice as its explanations may not be perceived as expected, especially with the lack of user-centric design of such models [21]. In addition, this method has been found to overlook the dependency between features in a model. It may also ask the user for unrealistic or unactionable changes, such as changing users' age or race [21, 23].

2.3 Existing User Segmentation Model

To develop an explainable model more effectively, the variety of users' backgrounds and goals must be considered. Thus, XAI researchers have applied the concept of user segmentation and defined robust segments that can be generalized to different applications, including social media. Most XAI studies use a similar segmenting basis, although the terminology of user groups may vary. They have examined users' explainability needs for AI decisions based on their backgrounds and technical expertise, classifying them into three categories: AI experts, domain experts, and lay users. Users fit into the first category if they have high technical expertise and may develop AI models. Domain experts are specialized in specific fields, such as healthcare, and use AI products to facilitate their decision-making. Whereas users who use AI products in their daily lives for entertainment, or information seeking and may lack technical expertise are the lay users [18, 25].

However, XAI techniques fall short in aligning explanations with the variety of users' needs. They mostly adopt a technical approach, so XAI scholars develop explanations by focusing on internal details of system modeling, such as feature importance or attention weights [3, 5, 9, 24]. These explanations are complex and cannot be interpretable by all users [9, 25]. Also, current techniques show a single explanation for all user types, assuming it fits all people. Although some social media applications may provide personalized explanations, they have been criticized for being vague, misleading, and incomplete, explaining only part of a model behavior [1]. They are also generated based on users' ad preferences or users' browsing history, without considering user categories [1, 9]. Thus, many researchers highlight the importance of user-centered design of XAI models in tailoring explanations to different user groups [9, 18, 20, 25]. Table 2 shows a comparison of the mentioned explainability methods and their suitability to the different user categories.

Table 2. XAI methods and their suitability to user categories

XAI method	Explanation features	User type	Level of suitability
SHAP	Consistent, Global & local feature contribution, Complex	Developer	High
		Domain experts	Moderate
LIME	Usable, Local feature contribution, Complex but can be simplified	Developer	High
		Domain experts	Moderate
		Lay users	Low
Counterfactual Explanations	Local feature contribution, No internal working details, What-if scenarios, User friendly	Domain experts	High
		Lay users	High

Legend: "High" indicates strong suitability for the given user group, based on clarity, detail, and user control. "Moderate" suggests partial alignment or interpretability. "Low" reflects limited usability due to complexity or abstraction

3 Vision and Proposed Framework

The purpose of this proposed framework is to develop a visualization tool that provides tailored explanations for AI recommendations, addressing diverse user needs in social media. The study aims to achieve the following objectives:

- Identifying the explainability needs of different social media users and contexts
- Designing a visualization prototype that can show context-specific explanations to the identified user categories and contexts
- Evaluating the tool by measuring its impact on users' decisions and trust in social media

The proposed system will be designed and evaluated progressively in three phases as follows:

- **Phase 1: User-type visualizer:** starting with technical versus non-technical users and running a user study using mock-up visual designs to explore their explanation preferences.
- **Phase 2: Context-aware extension:** identifying context-specific user scenarios based on a survey and mapping their needs to explanation formats.
- **Phase 3: Final visualization prototype and evaluation:** developing a functioning prototype with context-aware explanation options and running a user study to assess its impact on user trust and decisions.

Technical background is the segmentation basis to categorize users initially, so the tool starts to be framed by two major categories and their distinct needs. The category of domain experts, mentioned in Sect. 2.3, may not apply to this work as AI recommendations of social media are not industry-specific, and they do not target a specialized field. Also, domain experts themselves may need different explanations in different situations. Alternatively, context-specific scenarios will be defined in the second phase, focusing on what users need to understand in a specific situation rather than their profession-based perspectives. This is a central component of the proposed system, giving users a configurability option for different situations. User scenarios that reflect their intentions on social media can be defined, such as casual browsing, professional information gathering, and decision-making (on a product or service). The system tailors explanations to align with such scenarios, in addition to the alignment of user categories.

The system diagram is shown in Fig. 3, where user engagement data will be collected through social media APIs. The data will be used as input to the Amazon Personalize Platform, which is a fully managed machine learning service that is able to generate personalized recommendations. After generating social media like recommendations, they will be used to feed an explanation engine. This explainable model will then generate explanations tailored to user categories and contexts, which takes steps further than the current explainability practice highlighted in Table 1.

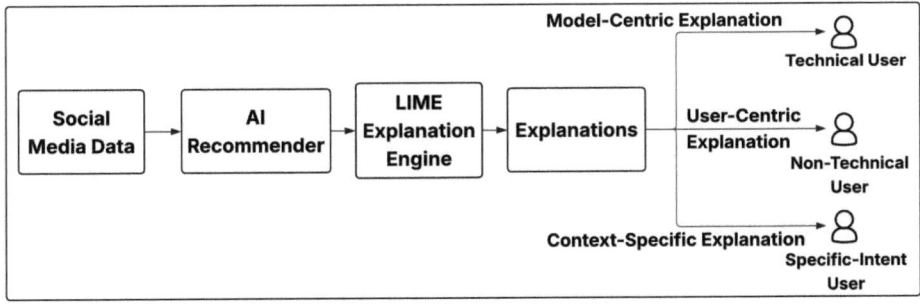

Fig. 3. System diagram

The vision of the proposed tool is to be user-friendly, showing understandable explanations to all users. In the context of social media, it is recommended to show hybrid explanations, combining multiple types of explanation formats to align with diverse user needs [8, 11]. These explanations usually include both model-centric and user-centric insights, and they can be represented in different ways, including visual and textual.

For this framework, we adopt a hybrid explanation approach that caters to both expert and casual audiences. For technically inclined users, the system displays a LIME-generated bar chart highlighting which features most influenced a recommendation. Non-technical users instead receive an icon-backed, plain-language rationale drawn from the same LIME output. A similar short explanation can be shown to users with casual browsing intent, while a comparison chart can be shown to users who want to decide on a product.

LIME is selected as an initial baseline for this framework due to its model-agnostic nature, ease of integration, and interpretable output format, as mentioned in Sect. 2.2. LIME's raw output (feature contributions) can be post-processed into formats that suit the user's goals, preferences, and level of technical expertise. Importantly, LIME generates local explanations, making it suitable for explaining individual AI recommendations within social media platforms, where recommendations are context-based and personalized. This aligns well with the tool's goal of delivering tailored, situational-specific, and user-friendly explanations. Additionally, given its simplicity and flexibility, it enables efficient prototyping (quickly building and testing early versions of an explainable tool) and facilitates the development of context-aware explanation interfaces. While LIME provides a practical foundation for this work, we also acknowledge the value of other XAI methods and highlight this as a direction for future comparative studies. As this is a vision paper, the current system is not yet fully implemented. We present a conceptual framework, mock-up designs, and planned evaluation methods, with system prototyping and deployment to follow in future work.

Figure 4 shows an example mock-up of visual explanations for different users. The top panel shows a simplified, icon-based explanation in plain language for lay users. The bottom panel presents a LIME-generated bar chart suited to technical users. Both visualizations are based on the same model output but adapted to different user needs. These visualization formats were designed to reflect the different explanation preferences identified in our user survey. While not directly tested, the contrasting formats aim to accommodate the needs of both non-technical and technical users as outlined in Sect. 4.

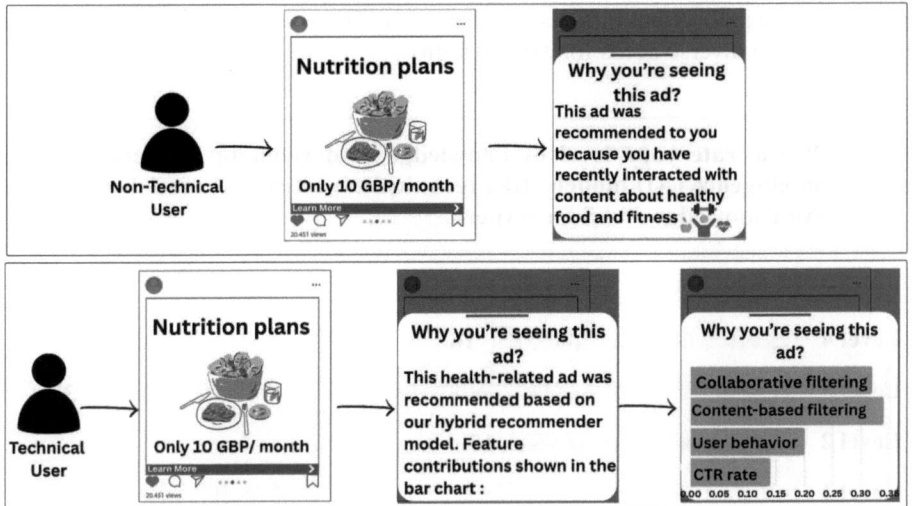

Fig. 4. Example mock-up of visual explanations for non-technical users vs. technical users

4 Case Example and Data Strategy

X (Twitter) will be selected as a case study for developing and testing the visualization tool. X has shown its obvious interest in improving its AI explainability and transparency and has collaborated with many experts and researchers. What makes this platform a suitable testbed for this work is that it is one of the few platforms where much of its content can be accessed through its API. Also, the nature of its content as short, textual, timestamped, and the availability of its metadata make it ideal for mimicking social media recommendations and improving an explainable AI visualization tool. In addition, Twitter is being used by diverse users with varying goals and backgrounds, offering a broad landscape to test user-specific explainability. Twitter has also been studied extensively in academic research [2, 6, 7], which can be supportive in data analysis, model evaluation, and benchmarking.

The data collection strategy for this study includes two parts. First, a user study will be conducted to explore the preferences of technical and non-technical users. The study will include surveys and interviews to recognize the situational needs of user context-specific scenarios. The second part of the data collection is fetching user engagement data from the X API. The data acquired using both methods complement one another, providing in-depth insights that will be considered to design the visualization tool.

As a starting point for understanding users' perspectives on social media explanations for AI-generated content, a survey was distributed among professionals specialized in different domains who are also social media users. The survey also establishes a foundation for user segmentation and identifying the distinct explainability needs. The survey includes a mix of binary, single-choice, rating scale, and short-answer questions, allowing a good balance of quantitative and qualitative data. We collected 106 responses from social media users with domain expertise across various fields. Participants were selected using a random sampling approach and invited through digital communication channels. These participants possess varying levels of technical knowledge, as shown in Fig. 5, reflecting diverse needs for explainability.

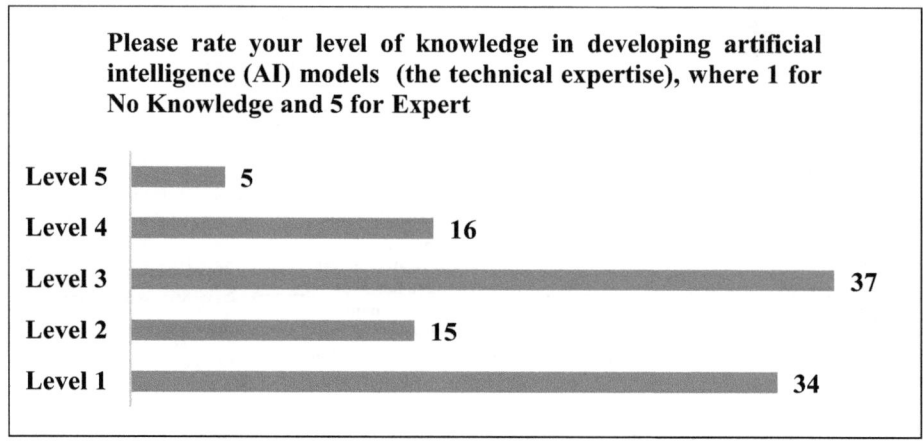

Fig. 5. Participants' levels of knowledge in AI

The collected data indicate the need for AI explainability in social media applications, as shown in Fig. 6. It is also clear that different users require different types of explanations. Figure 7 shows that 26% of respondents need detailed explanations, yet 50% prefer a simple explanation. When participants were asked about what would make explanations of social media recommendations more useful, one said: "If it is clear and simple…". Another participant said: "If it enhances my understanding of how those recommendations are made, as an AI expert". These results demonstrate the need for a configurable visualization explainability platform.

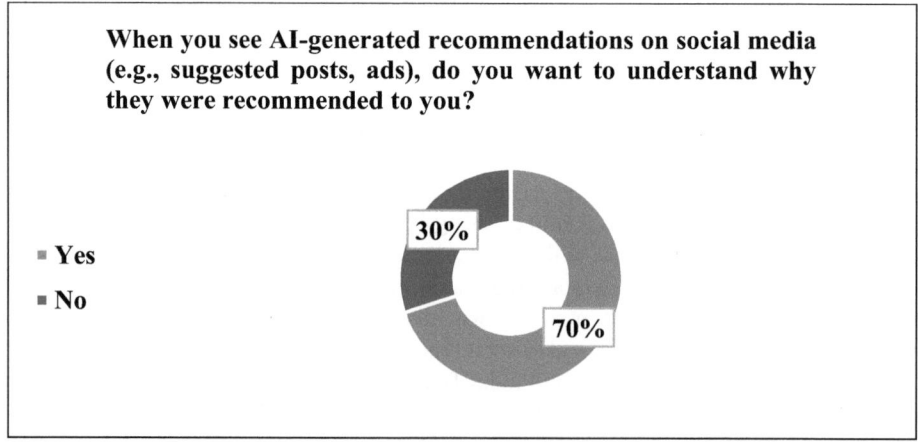

Fig. 6. Participants' need for explainability on social media

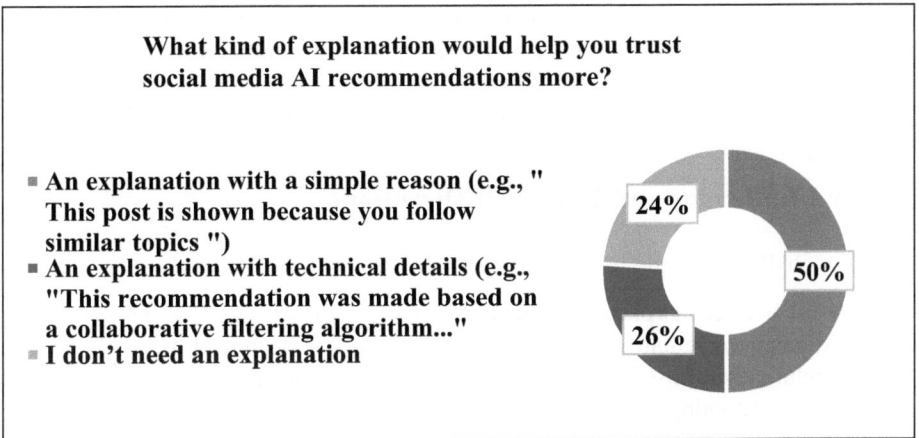

Fig. 7. Participants' preferences for social media explanations

The results also show that professionals from non-technology fields generally have low technical expertise. Also, 54% of the respondents agree that AI recommendations do not affect their professional decisions, as shown in Fig. 8. This outcome supports the current framework for segmenting social media users based on their technical background, not their fields of specialty. One participant mentioned: "Social media does not enter my professional space....".

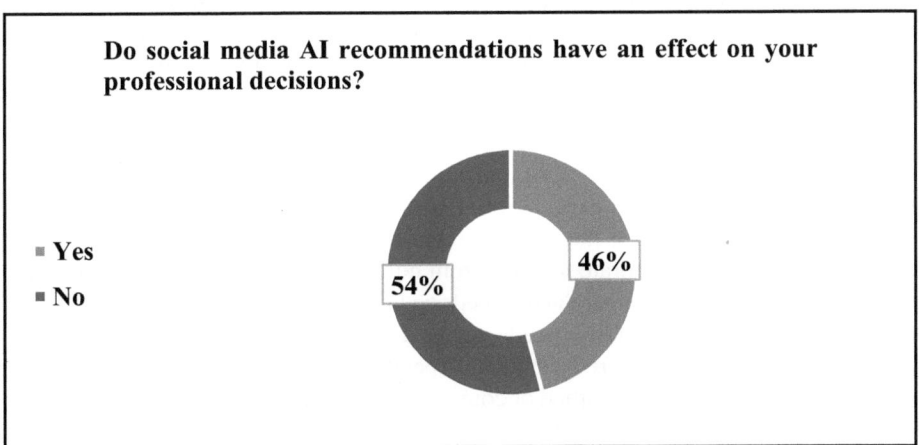

Fig. 8. Participants' perspectives on the impact of social media AI recommendations on their professional decisions

Generally, three key conclusions can be drawn from this survey: **1)** the desire for explainability by social media users, **2)** variations of explainability needs, and **3)** users' desire for more control and customizability. The overall results demonstrate a good motivation for developing a customizable visualization explainability tool for social media platforms. Additionally, the results validate that domain-specific explanations are generally unnecessary when using social media platforms, unless a specific context or situation is explicitly defined. Therefore, users' technical background should serve as the primary basis for segmenting social media users when designing an explainable tool.

5 Key Challenges and Research Directions

Three key challenges could be faced throughout the stages of this work:

1. **Trade-off between explanation simplicity and accuracy**: the issue lies in ensuring that explanations remain useful across all user types. A technical breakdown might overwhelm general users, yet oversimplified explanations risk eroding trust among specialists. Adopting a hybrid explanation approach, as mentioned in Sect. 3, can help to maintain a balance of both sides.
2. **API limitations:** potential access limitations of platform APIs, such as rate limits, request limits, data access restrictions, and constant policy updates, may affect the collection of the needed data. Using a local data storage, preprocessing, and prioritizing API requests may alleviate this problem. Alternatively, responsible web scraping can be considered with the retrieval of only public anonymous data to avoid ethical issues.
3. **Ethical and privacy considerations:** The use of social media data and user studies raises important ethical questions. As responsible data handling is key to building user trust and ensuring transparency, the system will ensure compliance with GDPR and institutional ethics policies. All participant data obtained through workshops, interviews, or surveys will be anonymized, securely stored, and collected with informed consent. For Twitter data, only publicly accessible information will be gathered, and all identifiable data elements will be anonymized.

While the proposed system is expected to provide promising results, improving social media explainability in terms of personalization and visualization based on user types, interactivity remains open for exploration. Future versions could include an interactive dashboard for users to select the desired explanation. In future iterations of our system, we also plan to incorporate insights from human-centered XAI literature to guide visual explanation design. Ribera and Lapedriza [18] highlight the importance of aligning explanation formats with user needs, goals, and expertise, suggesting, for example, that technical users benefit more from model-specific and abstract representations, while non-technical users prefer simplified or counterfactual explanations. Our ongoing development will use such guidance to refine the mapping between user groups and the visual formats introduced in Sect. 3. Additionally, we draw from Shneiderman's broader principles of human-centered AI [20], particularly the emphasis on explainability and trust through interactive and user-responsive interfaces. This aligns with our long-term goal of creating configurable visual explanations. Further research could also apply other XAI methods, such as SHAP or counterfactual explanations. These methods may offer other advantages, such as understanding cause-effect relationships (counterfactual explanations) or having more consistent and global explanations (SHAP). A comparative study can help assess their impact on other factors like explanation clarity and user satisfaction.

6 Conclusion

The current practice of social media explanations overlooks the diversity of user types and contexts. This vision paper presents a context-specific explainable system that offers tailored visual reasoning to users based on their varying needs. Future work will focus on validating the proposed framework and evaluating its impact on trust and engagement by conducting surveys, user studies, data analysis, and going through prototype development and final assessment. We also plan to extend the framework to include additional explanation modes, such as interactive sliders for explanation depth and voice-based explanations for accessibility. Furthermore, future iterations will focus more on visual language approaches as well as investigating longitudinal effects of explanation quality on user retention, trust calibration, and resistance to algorithmic bias in social media.

Disclosure of Interests. The authors have no competing interests to declare that are relevant to the content of this paper.

References

1. Andreou, A., et al.: Investigating ad transparency mechanisms in social media: a case study of facebook's explanations. In: Proceedings 2018 Network and Distributed System Security Symposium. Internet Society, San Diego, CA (2018). https://doi.org/10.14722/ndss.2018.23191
2. Andryani, R., et al.: Social media analytics: data utilization of social media for research. Journalisi **1**, 2, 193–205 (2019). https://doi.org/10.33557/journalisi.v1i2.23
3. Benabdelouahed, R., Dakouan, C.: The use of artificial intelligence in social media: opportunities and perspectives. Expert J. Market. **8**(1), 82–87 (2020)
4. Dieber, J., Kirrane, S.: Why model why? Assessing the strengths and limitations of LIME (2020). https://doi.org/10.48550/ARXIV.2012.00093
5. Ehsan, U., et al.: Human-Centered Explainable AI (HCXAI): beyond opening the black-box of AI. In: CHI Conference on Human Factors in Computing Systems Extended Abstracts, pp. 1–7. ACM, New Orleans LA USA (2022). https://doi.org/10.1145/3491101.3503727
6. Fiok, K., et al.: Analysis of sentiment in tweets addressed to a single domain-specific Twitter account: comparison of model performance and explainability of predictions. Expert Syst. Appl. **186**, 115771 (2021). https://doi.org/10.1016/j.eswa.2021.115771
7. Guesmi, M., et al.: On-demand personalized explanation for transparent recommendation. In: Adjunct Proceedings of the 29th ACM Conference on User Modeling, Adaptation and Personalization, pp. 246–252. ACM, Utrecht Netherlands (2021). https://doi.org/10.1145/3450614.3464479
8. Haque, A.B., et al.: Explainable Artificial Intelligence (XAI) from a user perspective: a synthesis of prior literature and problematizing avenues for future research. Technol. Forecast. Soc. Chang. **186**, 122120 (2023). https://doi.org/10.1016/j.techfore.2022.122120
9. Haque, A.B., et al.: To explain or not to explain: an empirical investigation of AI-based recommendations on social media platforms. Electron Mark. **35**, 1, 2 (2025). https://doi.org/10.1007/s12525-024-00741-z
10. Imbwaga, J.L., et al.: Explainable hate speech detection using LIME. Int. J. Speech Technol. **27**(3), 793–815 (2024). https://doi.org/10.1007/s10772-024-10135-3

11. Kouki, P., et al.: Generating and understanding personalized explanations in hybrid recommender systems. ACM Trans. Interact. Intell. Syst. **10**(4), 1–40 (2020). https://doi.org/10.1145/3365843
12. Mehta, H., Passi, K.: social media hate speech detection using Explainable Artificial Intelligence (XAI). Algorithms **15**, 8, 291 (2022). https://doi.org/10.3390/a15080291
13. Molnar, C.: Interpretable machine learning: a guide for making black box models explainable. Christoph Molnar, Munich, Germany (2022)
14. Molnar, C.: Interpreting machine learning models with SHAP: a guide with Python examples and theory on shapley values. Chistoph Molnar c/o MUCBOOK, Heidi Seibold, München, Germany (2023)
15. Noah, O.: Evaluating explainability in AI models: comparing SHAP, LIME, and other techniques (2024)
16. Nohara, Y., et al.: Explanation of machine learning models using shapley additive explanation and application for real data in hospital. Comput. Methods Programs Biomed. **214**, 106584 (2022). https://doi.org/10.1016/j.cmpb.2021.106584
17. Ozmen Garibay, O., et al.: Six human-centered artificial intelligence grand challenges. Int. J. Hum.–Comput. Interact. **39**, 3, 391–437 (2023). https://doi.org/10.1080/10447318.2022.2153320
18. Ribera, M., Lapedriza, A.: Can we do better explanations? A proposal of user-centered explainable AI. Los Angeles (2019)
19. Sadiku, M.N.O., et al.: Artificial intelligence in social media. Int. J. Sci. Adv. **2**, 1 (2021). https://doi.org/10.51542/ijscia.v2i1.4
20. Shneiderman, B.: Human-centered artificial intelligence: reliable, safe & trustworthy (2020). https://doi.org/10.48550/arXiv.2002.04087
21. Spreitzer, N., Haned, H.: Evaluating the practicality of counterfactual explanations (2022)
22. Verma, S., et al.: Counterfactual explanations and algorithmic recourses for machine learning: a review (2020). https://doi.org/10.48550/ARXIV.2010.10596
23. Verma, S., et al.: Counterfactual explanations for machine learning: challenges revisited (2021). https://doi.org/10.48550/ARXIV.2106.07756
24. Vilamala, M.R.: Visualizing logic explanations for social media moderation. In: International Joint Conference on Autonomous Agents and Multiagent Systems, vol. 2023, pp. 3056–3058 (2023)
25. Yang, L. et al.: What does it mean to explain? A user-centered study on AI explainability. In: Degen, H., Ntoa, S. (eds.) Artificial Intelligence in HCI, pp. 107–121. Springer International Publishing, Cham (2021). https://doi.org/10.1007/978-3-030-77772-2_8

Transparent Adaptive Learning via Data-Centric Multimodal Explainable AI

Maryam Mosleh(✉), Marie Devlin, and Ellis Solaiman

School of Computing, Newcastle University, Newcastle Upon Tyne NE1 7RU, UK
{m.mosleh1,marie.devlin,ellis.solaiman}@newcastle.ac.uk

Abstract. Artificial intelligence-driven adaptive learning systems are reshaping education through data-driven adaptation of learning experiences. Yet many of these systems lack transparency, offering limited insight into how decisions are made. Most explainable AI (XAI) techniques focus on technical outputs but neglect user roles and comprehension. This paper proposes a hybrid framework that integrates traditional XAI techniques with generative AI models and user personalisation to generate multimodal, personalised explanations tailored to user needs. We redefine explainability as a dynamic communication process tailored to user roles and learning goals. We outline the framework's design, key XAI limitations in education, and research directions on accuracy, fairness, and personalisation. Our aim is to move towards explainable AI that enhances transparency while supporting user-centred experiences.

Keywords: Explainable Artificial Intelligence (XAI) · Adaptive Learning Systems · Human-Centred AI · Generative AI · Multimodal Explanations · AI in Education

1 Introduction

AI and personalised learning have driven the development of adaptive learning systems. While these systems have made significant strides in tailoring content to learners, the explanations behind AI-driven decisions remain opaque and generic [18].

Adaptive learning systems adapt to learners' performance and preferences by constantly tailoring the education style and tasks based on insights derived from learner engagement with the learning material [16, 20]. Data is collected and analysed through various AI and data analytics tools, including machine learning, Bayesian networks, neural networks, and educational data mining [10]. Despite personalised content delivery, these systems often lack transparency [6]. The rationale behind content selection and learner assessment remains unclear, creating a 'black box' effect. This challenge can undermine trust among learners and educators and negatively impact their engagement when they're unsure of the system's validity and relevance to their learning journey [27]. While current XAI techniques use textual and visual explanations, most adaptive learning systems mainly focus on only providing non-personalised text explanations and rarely use personalised visual aids [18]. This limitation can negatively impact the effectiveness of AI decision explanations for learners with diverse preferences and needs [29]. While visual explanation techniques such as heatmaps and saliency maps are some of the most used visualisation methods in XAI, the general use of visual explanations in education remains limited [31].

In response to these limitations, it is essential to integrate personalised and context-specific XAI approaches into AI-driven adaptive learning systems. This paper presents a hybrid, user-centric explainability framework that uses traditional XAI methods and generative AI to create personalised, multimodal explanations tailored to the needs of diverse educational stakeholders. The key contributions of this paper are the following: 1) a review of current XAI limitations in education; 2) a novel hybrid framework combining traditional and generative XAI; 3) a conceptual pipeline for user-specific explanation delivery; and 4) a roadmap for operationalising personalised explainability. The remainder of this paper is organised as follows: Sect. 2 reviews current XAI techniques and their educational limitations. Section 3 introduces our proposed hybrid framework, supported by a conceptual architecture. Section 4 outlines open challenges and research directions. Section 5 discusses the potential impact, and Sect. 6 concludes with future work.

2 Limitations of Current XAI in Education

Despite XAI techniques' progress in recent years, their integration in adaptive learning systems has been limited. Unfortunately, most of these XAI techniques, including SHAP, LIME, and counterfactual explanations, rely on algorithmic interpretability, hence static and non-personalised outputs. They follow a "one-size-fits-all" approach that provides explanations that don't cater to the preferences of different users, making AI transparency less inclusive [3].

Moreover, these techniques rarely consider the distinct needs of various user groups, such as students, instructors, administrators, and other stakeholders [18]. Each may require different levels of depth and presentation formats in AI explanations. Failing to address these diverse requirements can limit transparency and reduce trust in adaptive learning systems.

2.1 Common XAI Techniques and Their Educational Constraints

This section reviews widely used XAI techniques and highlights their educational limitations, focusing on how each method supports (or fails to support) personalisation, clarity, and multimodal delivery.

SHAP (SHapley Additive exPlanations) is an explainability method based on Shapley values from cooperative game theory. It assigns the importance of each feature in a machine learning model's prediction, indicating its impact on the model's final decision, where it provides consistent and mathematically validated explanations [21]. SHAP is widely adopted in financial institutions to interpret complex machine learning models, especially in credit scoring helping lenders make well-informed decisions [9]. However, learners without specialised knowledge may struggle to understand such outputs.

LIME (Local Interpretable Model-Agnostic Explanations): LIME uses an approximation technique that explains AI decisions by generating a local interpretable model around a given instance. It alters the input data to observe the shift in predictions, making it clearer to understand the process of AI decisions [22]. However, methods like SHAP and LIME often assume technical understanding and produce static explanations ill-suited for dynamic, user-focused educational settings. Also, the outputs tend to be sensitive to small changes in data, which introduces inconsistent interpretations. Moreover, LIME's effectiveness is further restricted due to the difficulty of handling high-dimensional, sequential or multimodal educational datasets, which limits its usefulness for complex learner paths.

By providing "what if" scenarios, **Counterfactual explanations** will present how slight alterations in the input data can affect the final outcome, helping learners understand the boundaries and the basis on which the model decisions are based [11]. Similarly, counterfactual explanations may suggest unrealistic interventions unless grounded in educational logic. However, they can propose unfeasible or educationally unsuitable interventions without careful adherence to pedagogical logic and guidance by education logic.

While SHAP, LIME, and Counterfactual explanations are well-known XAI techniques, a broader landscape of methods may be particularly valuable in educational settings. Approaches like: **Anchor** intuitively generates high-precision decision rules [28]. Although these rules enhance interpretability beyond what feature attribution techniques offer, Anchor fails to capture the learning process's actual complexity and non-linear nature. **Surrogate Decision Trees** mimic complex models with simpler, interpretable decision trees [12]. Despite their clarity, they may overlook subtle model behaviours and interactions that are particularly important to consider in adaptive learning systems. For example, learners' prior knowledge, engagement level, and response time can all impact the system's content suggestions.

Gradient-based visual explanation methods (e.g., saliency maps or Grad-CAM) [30] can offer clear insights into model reasoning by highlighting the areas in the input that significantly influence the model's decisions. While these techniques perform well in visual tasks (e.g., image-based emotion recognition), they're less effective with textual education data.

To develop effective and inclusive AI explainability for adaptive learning, we must investigate how these techniques can be tailored to varied educational needs and contexts. This exploration involves evaluating each method's interpretability, computational feasibility, and user preference, ensuring that the final system supports a wide range of learner profiles.

2.2 Comparison of XAI Techniques in Education

See Table 1.

Table 1. Comparison of XAI Techniques in Educational Contexts.

Method	Strengths	Weaknesses in Education	Level of Personalisation	Visual/Text Capabilities
SHAP	Strong theoretical foundation	Too technical for most users	High- It can be tailored with learner data	Graphs and textual explanations
LIME	Straightforward and widely compatible	Struggles with complex or mixed formats	Low – static explanations	Basic text and plots
Counterfactuals	Learner based improvement cues	May suggest unrealistic actions	High – tailored via learner input	Text-based, with visual option
Anchors	Generates accurate understandable explanations	May oversimplify user needs	Medium - customisable, but not adaptive	Primarily text-based explanations
Surrogate Decision Trees	Clear via if-then-else rules	Poor fit for complex, non-linear systems	Medium – limited personal relevance	Visual trees and textual explanations
Gradient Visualisation Methods	Real-time insight into key factors	Limited causal insight	Low – not learner-specific	Visual outputs (e.g., heatmaps)

2.3 Common XAI Techniques and Their Educational Summary of Gaps

For XAI techniques to be effective in educational settings, they require more than generic feature attribution; they must offer context-specific, learner-friendly and educationally meaningful explanations.

However, current existing methods generally fail to address the following criteria:

(1) **Contextual interpretability:** Explanations must be personalised based on different users' backgrounds and learning paths [17].
(2) **Multimodal integration:** The need to handle different data types, both visual and textual.
(3) **Transparency:** Enabling users to understand and influence the reasoning behind AI's decisions. Despite their potential, current XAI techniques rarely address the varied interpretability needs of educational stakeholders. Their limitations in delivering adaptive, multimodal explanations underline the need for a new framework that places user context at its core.

2.4 Common Adaptive Learning Systems and Their Educational Summary of Gaps

Table 2. Comparison of Adaptive Learning Systems with the proposed framework

Feature/Aspect	AutoTutor	GnuTutor	ALEKS	Knewton	Our Framework
Focus	STEM tutoring with conversation	Replicating AutoTutor	Mastery learning in STEM	Adaptive content	Personalized explainable AI
Explainability	Moderate: Basic emotion and scripts	Low: Fixed, scripted responses	Low: Outcome-only focus	Low: black-box models	High: Adaptive, multimodal feedback
Personalization	Minimal: No persistent learner model	Minimal: (template-driven)	Medium: Adaptive knowledge model	High (data-driven predictions)	High (context-aware, preference-based)
Collected Data Types	Typed responses, timing, emotion (limited)	Typed responses, interaction logs.	Answers to concept-specific problems	Usage and performance data	Interaction data, user preferences, role context
Target Users	Learners only	Educators, Learners, researchers, and developers	K–12 and higher education	Higher education	Student, teachers, admins

As shown in Table 2, AutoTutor offers real-time adaptation through Latent Semantic Analysis (LSA), cognitive state tracking and emotion detection, it is ultimately constrained by a finite-state model built on pre-scripted responses and limited set of instructional moves (e.g., prompts, hints, affirmations), limiting adaptation to scripted sequences and shallow interpretation of context. The selected instructional moves are only guided by immediate user input and surface-level emotional cues. There is no generative feedback, adaptation to learning profiles, or tailoring of teaching strategies based on users' preferences [23]. GnuTutor is an open-source implementation that replicates AutoTutor's functional elements, such as LSA-based semantic analysis, speech act classification, mixed-initiative dialogue, and animated agents, while removing licensing barriers and providing a simpler way of deployment [24]. Although Gnututor's prolog (a logic-based language, used in AI and dialogue systems [25]) based architecture provides a more accessible codebase, like AutoTutor, it is still bound to finite-state and script-driven interactions, lacking the flexibility to tailor instructions based on the user's ongoing needs and preferences [26]. As a result, personalisation remains limited and reactive (triggered by immediate learner input, including behavioural or emotional cues). It lacks a mechanism for continuous user modelling, and thus cannot adjust instructional formats, such as depth, based on prior learner interactions or preferences [5].

While AutoTutor and GnuTutor can excel in delivering structured approaches to tutoring, they still share the same limitations of static personalisation, script-bound interaction, a lack of generative adaptability to user's preferences and needs, and the lack of transparent reasoning processes as they don't provide explanations or personalisation features that our proposed framework does. By integrating XAI, generative models and real-time user data modelling, our system supports broader functionality across different user roles.

While Autotutor and GnuTutor rely on structured pedagogical dialogue and pre-scripted responses, Knewton adopts a data-centric approach that integrates psychometric profiling, collecting and analysing users' data using AI to estimate their skills, preferences, needs, and knowledge [32]. Furthermore, it utilises content graph alignment, where it structures the learning material in the form of knowledge graph that incudes aligning interconnected concepts with learner's current level of understanding and performance, guiding the system into the next best step and concept based on what the learner has improved in and what's connected to it [2, 32]. Despite offering performance metrics and visual dashboards, the underlying logic behind the adaptive decisions remains inaccessible, raising concerns regarding the reliability of the system's automated interventions [14].

Although some adaptive learning systems provide explanations for AI decisions, the AI models used often depend on pre-set rules and parameters, resulting in static and non-personalised explanations. This inflexibility could negatively impact the user's experience as the explanations do not account for their different needs and preferences [29]. For example, the instructor's manual of the Assessment and Learning in Knowledge Spaces (ALEKS) system includes *"explanations and algorithmically generated practice problems"*, stating that the explanations of the material and AI decisions are not tailored to the user's needs and preferences [1]. Instead, ALEKS provides standardised explanations for all users, following a "one-size-fits-all" approach.

The study by Conati et al. [7] on tutoring systems illustrates how personalised explanations of AI decisions can improve learners' trust and engagement with these systems, emphasising the importance of designing future learning systems that don't only focus on delivering accurate content but also provide different explanation techniques that adapt to learners unique profiles, hence adopting a human-centric perspective. This aligns with our proposed framework's aim to provide personalised, multimodal explanations, promote transparency with AI decisions, and improve users' trust, thus closing the gap between the AI decision-making process and users' understanding.

3 Proposed Framework

Despite existing research on personalisation, current XAI techniques fail to deliver meaningful, user-specific explanations. This creates a disconnect between adaptive systems' potential and user experience. Although adaptive learning systems can adjust learning content based on users' progress, the provided explanations are often static and generic, making it difficult for learners to understand the reasoning behind AI recommendations. The rationale for combining generative AI with traditional XAI is grounded in the challenge that most raw explanations from SHAP, LIME, or counterfactual methods are either overly technical or not adapted to learner roles. Generative AI models have shown promise in translating structured data into natural language that matches user comprehension levels (e.g., OpenAI's GPT-4o use in education as shown in Kim et al.'s (2024) study [19], where it was used to tailor scientific information to individual learner profiles and shown its effectiveness in improving user's understanding). We extend this idea by using generative models as a translation layer for XAI outputs. We propose a hybrid explainability framework to address this gap, which will generate adaptive, multimodal explanations tailored to user roles and preferences.

3.1 Overview of the Hybrid Framework

The framework design includes four main stages:

1. **Data Collection and Learner Profiles:** We begin by analysing existing XAI methods used in adaptive learning systems to identify their current limitations. In parallel, we will categorise educational stakeholders (such as students, teachers, module leaders, and administrators), and determine their specific explainability needs through interviews, focus groups, and surveys. This process enables the creation of dynamic learner profiles that accurately reflect user knowledge, goals, and contexts.
2. **AI Decision Engine and XAI Layer:** Learner data will be processed by an adaptive learning system using models such as Bayesian Knowledge Tracing (BKT) [4] to tailor educational content. The decisions generated by these models will be interpreted via a dedicated XAI layer, which selects and applies the most suitable explainability method according to each user's profile and preferences.
3. **User-friendly explanations through Generative AI:** Generative AI will convert technical XAI outputs into accessible, conversational explanations. For example, rather than presenting a technical explanation like: *"SHAP value of -0.3 for concept node algebraic expressions"* the system would generate a more user-friendly message such as: *"We noticed that you spent extra time solving the last two algebraic expression problems, so we are offering additional practice to improve your understanding."*
4. **Personalisation:** Explanation delivery will be tailored to each user type and context:
   (a) Students will receive simple, motivational explanations in both text and visual.
   (b) Teachers will access detailed dashboards showing student progress, knowledge gaps, and performance-enhancing suggestions.
   (c) Administrators will be provided with high-level system trend summaries and user engagement reports.

This framework therefore aims to produce explainable AI, and also deliver explainable-to-the-user AI. Explanations are personalised in language style, delivery format, and depth, aligning with individual user roles, preferences, and cognitive needs.

3.2 Conceptual Pipeline Diagram Description

Designed for adaptive learning environments, the suggested framework introduces a layered conceptual pipeline to support delivering personalised, multimodal explanations of AI decisions in these systems. Figure 1 presents the overall architecture of our proposed

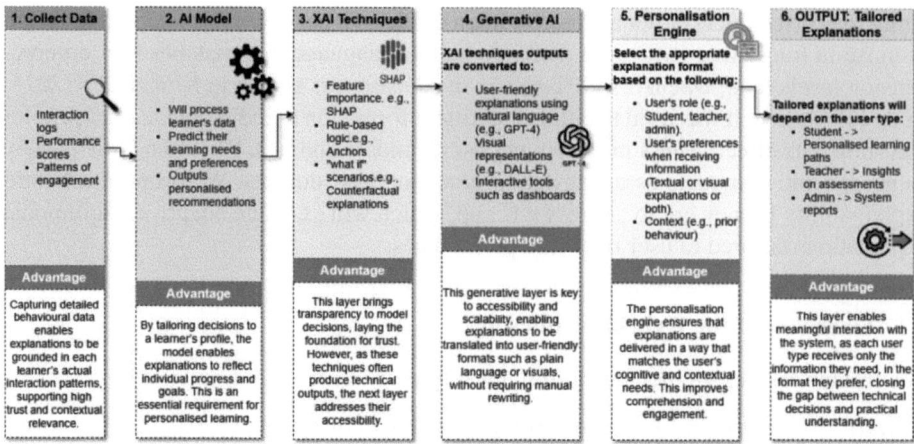

Fig. 1. Conceptual Architecture of the Proposed Hybrid Explainability Framework.

hybrid explainability framework. It consists of six layers: data collection, AI decision-making, XAI explanation generation, generative AI translation, a personalisation engine, and delivery of tailored outputs. This layered approach ensures that explanations are technically accurate, context-aware, user-friendly, and aligned with individual learning roles and needs. Each layer is described in detail below.

Layer 1: Collect Data: The system collects continuous learner data (performance, engagement) to inform decisions. This provides the AI model with the rich data needed to understand each user's different learning paths.

Layer 2: AI Model: The AI model then uses the collected data to generate personalised recommendations and predictions, such as suitable learning materials and potential areas of misunderstanding. However, the decision's lack of explanation still challenges learners and educators to understand or trust it fully.

Layer 3: XAI Techniques: To tackle this problem and clarify how AI decisions are made, this framework will integrate different XAI techniques that will help uncover the reasoning behind AI decisions by emphasising the key features affecting its outputs.

Layer 4: Generative AI: Though technically valid, XAI outputs are often complex. Generative AI simplifies them into accessible text or visuals.

Layer 5: Personalisation Engine: Is the decision-making layer. By analysing user roles, preferences, and interaction context, it will dynamically select the most appropriate explanation method, format, and depth for each individual. This layer includes: (1) user profile identification (e.g., student, teacher, admin); (2) contextual analysis to align content and explanations with user goals; (3) format selection based on profile and context; and (4) a feedback loop that updates profiles from ongoing engagement.

Layer 6: OUTPUT: Tailored Explanations: The final personalised explanations are delivered at this layer with a suitable depth and format based on the user's requirements and preferences. For example, learners will receive adaptive feedback tied to their progress, instructors or teachers will receive specific insights to inform their teaching decisions, and administrators will receive high-level performance monitoring reports.

3.3 Illustrative Examples of Personalised Explanations

To illustrate how our framework operates, we present hypothetical examples designed to reflect the needs of different user types:

- **Student**: "You've done well on basic recursion problems but got stuck when it involved trees. That's common—tree problems are harder because they involve multiple recursive calls. Here's a step-by-step example to help you practice breaking it down."
- **Teacher**: "Several students in your class struggled with dynamic programming this week. Most made the same mistake: not storing previous results. A short group review using visual aids on memoization might help reinforce the core idea."
- **Administrator**: "Data shows that first-year students in the online program are spending significantly more time on introductory algorithm modules compared to their in-person peers. This could indicate a need for additional scaffolding or better pacing online."

4 Key Challenges and Open Questions

While the vision for Human-Centric, Multimodal Explainable AI in adaptive learning systems holds immense potential, it also raises foundational questions and risks. Instead of offering definite answers, this section seeks to highlight these questions and risks that could guide cross-disciplinary investigation.

4.1 Accurate, Faithful and Personalised Explanations

Question: *How can personalised explanations remain accurate and faithful to the model?*

Personalised explanations are intended to adapt to different user roles and cognitive needs; however, they must also accurately reflect the model's underlying logic and reasoning. The challenge lies in preserving the fidelity of AI decisions while adapting their presentation to suit learners, teachers, or administrators. The accuracy and relevance of these personalised outputs may not always align with established standards of explainability, such as user transparency, as highlighted in prior work, including EU AI regulation [8]. This raises concerns about whether tailoring explanations for each user might inadvertently introduce bias, misrepresentation, or inconsistency. Techniques that validate the educational and logical integrity of personalised outputs will be essential to ensure both trust and utility. To address this within our research, the system will include validation mechanisms ensuring that generated explanations faithfully reflect the underlying model's reasoning, verified through expert review and alignment with model outputs.

4.2 Understandability, Accuracy, and Fairness

Question: *What trade-offs exist between understandability, accuracy, and fairness?*

Even when fidelity to the model is preserved, adaptive learning systems must still address the inherent tension between explanation clarity and technical detail. Over simplifying explanations to improve accessibility may reduce their usefulness or lead to misunderstandings, while maintaining full technical accuracy may make them incomprehensible to many users. In educational contexts, these trade-offs can influence learner development, engagement, and trust in AI systems. Prior work by Holstein et al. [13] also stresses the importance of fairness and transparency in AI systems. Solutions require testing with users to balance clarity, accuracy, and fairness. In our work, this trade-off will be empirically studied in user trials, where multiple explanation types (e.g., simple vs. detailed) will be compared across different learner profiles to identify optimal balances.

4.3 Generative AI in Education

Question: *Can generative AI be reliably used to explain critical decisions in education?*

Recent advances in large language models (such as ChatGPT) have introduced powerful new capabilities for delivering conversational and adaptable explanations and supporting interactive learning. However, these models still raise serious concerns regarding the accuracy of the information provided, including the reasoning behind the model's decisions. Generative AI may generate biased, irrelevant, or inaccurate results. Appropriate control mechanisms will be implemented to address this, and the model will be fine-tuned using a diverse dataset, followed by iterative testing and refinement. Furthermore, multiple validation layers will be required to evaluate the reliability of the generated explanations. This includes evaluations from experts in the education sector to verify the pedagogical accuracy. At the same time, students' feedback that will be gathered through user studies in low-risk and controlled settings (e.g., formative assessments) can assess the interpretability of the generated explanations and their usefulness. To ensure reliable generative outputs, models will be fine-tuned on educational datasets and constrained through templates aligned with AI decisions, combined with human validation.

4.4 Adapting to User Explanation Preferences

Question: *How should explanation preferences be modelled and updated for different users?*

Every user has unique preferences, needs, and requirements; some learners may prefer visual explanations, while others favour textual explanations or detailed, structured guidance. These preferences may change over time as users become more skilled, presenting the need for dynamic adaptation within learning systems. User preferences will be initially captured through onboarding surveys and refined through interaction data (e.g., skipping visuals), with dynamic updates via a feedback loop.

5 Potential Impact

The proposed framework has the potential to significantly enhance the interaction experience between learners and education technologies. Delivering personalised and user-friendly explanations in real time helps address the critical challenges of AI-driven adaptive learning systems and educational AI by ensuring clarity and fairness.

Potential Impact on Students:

- Enhance system's transparency, hence improving users' trust in AI decisions and engagement with learning systems.
- Promoting metacognitive awareness, encouraging users to self-monitor and evaluate their progress on their learning journey [15].
- Users' roles will change from passive consumers to collaborators in their learning process.

Potential Impact on Teachers:

- Gain deep and transparent reports on student engagement and learning progress.
- Traditional performance metrics will be replaced with personalised explanations, allowing teachers to see the adaptive logic driving AI decisions.
- Enhance the interaction between users and AI, supporting a data-driven and responsive teaching practice.

Potential Impact on Institutions:

- This framework will ensure the main criteria of responsible AI, accountability, fairness, transparency, and adherence to institutional and legal guidelines, as institutions can align algorithmic behaviour with their ethical standards, domain-specific benchmarks, and accreditation bodies' expectations.
- It can help institutions achieve their educational goals and enhance student development through early intervention. It ensures that the decisions and recommendations it provides align with the set goals, such as improving learning outcomes and supporting student reflection and confidence building ahead of summative assessments.
- Explainable AI decisions support external validation, where parents can easily understand their child's learning path and progress, and policymakers can easily evaluate the system.

Potential Impact on Other Domains: Healthcare, Finance and Social Platforms: Although education remains the primary focus of this paper, the core principles of this framework can be applied to other high-stakes sectors where trust and human-AI interaction are critical parts of them. In the healthcare sector, personalised explanations could aid patients' understanding of treatments. In finance, personalised rationales for decisions that depend on algorithms, such as credit scoring, could support users to make an informed decision. In social platforms, adaptive explanations can provide the clarity needed to understand certain content choices, including filtering or prioritising.

6 Conclusion

We presented a vision for hybrid explainability in adaptive learning systems that integrates user modelling, traditional XAI, and generative AI. This human-centric approach aims to enhance transparency, comprehension, and trust. In the next phase, we will conduct an extensive literature review to discover and evaluate the current AI-driven adaptive learning systems and identify their limitations in explaining AI decisions. This will then be followed by conducting user studies within Newcastle University's School of Computing to assess the impact of explanation types on trust and engagement. Methods such as surveys, focus groups, and semi-structured interviews will be employed to understand what each group considers a 'meaningful' explanation and how they prefer to receive it. Findings from this phase will guide the selection or development of XAI techniques and the generation of personalised explanations. Next is systematically evaluating a range of XAI and how Generative AI can complement these methods. The system will be tested with Newcastle University School of Computing students (e.g., Students from a specific module within the School of Computing) and educators. Data will be collected via workshops, interviews, and questionnaires to assess trust, understanding, and engagement. Finally, a controlled study will be conducted with multiple experimental groups, each exposed to different forms of AI explanations (e.g., numeric-based, rule-based, text-based, visual, and hybrid). Participants' trust, understanding, and engagement levels will be measured. We will also analyse whether certain user groups (e.g., novices vs. advanced learners, teachers vs. admins) prefer specific explanation types. This paper outlines a forward-looking vision for hybrid explainability in adaptive learning. Future work will implement and evaluate this framework through participatory design with learners and educators.

References

1. ALEKS Corporation: ALEKS Instructor's Manual for Math Prep for Accounting. ALEKS Corporation (2013)
2. Alta: Blog-What are Knewtons Knowledge Graphs. https://www.wiley.com/en-us/education/alta/resources/blog-what-are-knewtons-knowledge-graphs. Accessed 3 July 2025
3. Bove, C., et al.: Why do explanations fail? A typology and discussion on failures in XAI. https://arxiv.org/pdf/2405.13474 (2024). https://doi.org/10.48550/arXiv.2405.13474
4. Bulut, O., et al.: An introduction to bayesian knowledge tracing with pyBKT. MDPI Open Access J. (2023). https://doi.org/10.3390/psych5030050
5. C. Graesser, A., et al.: Intelligent tutoring systems. In: Intelligent Tutoring. University of Memphis, Memphis
6. Cheong, B.C.: Transparency and accountability in AI systems: safeguarding wellbeing in the age of algorithmic decision-making, vol. 6, p. 11 (2024). https://doi.org/10.3389/fhumd.2024.1421273
7. Conati, C., et al.: Toward personalized XAI: a case study in intelligent tutoring systems. https://arxiv.org/abs/1912.04464 (2019). https://doi.org/10.48550/arXiv.1912.04464

8. European Commission: Proposal for a Regulation of the European Parliament and of the Council Laying Down Harmonised Rules on Artificial Intelligence (Artificial Intelligence Act) and Amending Certain Union Legislative Acts (2021)
9. Fathima, S.: LIME vs SHAP: a comparative analysis of interpretability tools. https://www.markovml.com/blog/lime-vs-shap.
10. Gligorea, I., et al.: Adaptive learning using artificial intelligence in e-learning: a literature review, vol. 13, p. 12 (2023). https://doi.org/10.3390/educsci13121211
11. Guidotti, R.: Counterfactual explanations and how to find them: literature review and benchmarking. Springer Nat. Link. **38**, 2770–2824 (2022). https://doi.org/10.1007/s10618-022-00831-6
12. Herbinger, J., et al.: Leveraging model-based trees as interpretable surrogate models for model distillation. https://arxiv.org/abs/2310.03112? (2023). https://doi.org/10.48550/arXiv.2310.03112
13. Holstein, K., et al.: Improving fairness in machine learning systems: what do industry practitioners need? Presented at the January 7 (2019). https://doi.org/10.48550/arXiv.1812.05239
14. innovator, Z. the: Lack of Explainability and Transparency in AI: The Black-Box Dilemma. https://medium.com/@ThisIsMeIn360VR/lack-of-explainability-and-transparency-in-ai-the-black-box-dilemma-5bb58776cd93. Accessed 3 July 2025
15. Isaacson, R.M., Fujita, F.: Metacognitive knowledge monitoring and self-regulated learning: academic success and reflections on learning, vol. 6, pp. 39–55 (2006)
16. Kabudi, T., et al.: AI-enabled adaptive learning systems: a systematic mapping of the literature. Comput. Educ. Artif. Intell. **2** (2021). https://doi.org/10.1016/j.caeai.2021.100017
17. Khan, M.A., Khojah, M.: Artificial intelligence and big data: the advent of new pedagogy in the adaptive e-learning system in the higher educational institutions of Saudi Arabia. Educ. Res. Int. (2022). https://doi.org/10.1155/2022/1263555
18. Khosravi, H., et al.: Explainable artificial intelligence in education. Comput. Educ. Artif. Intell. **3** (2022). https://doi.org/10.1016/j.caeai.2022.100074
19. Kim, T., et al.: Steering AI-driven personalization of scientific text for general audiences, http://arxiv.org/abs/2411.09969 (2024). https://doi.org/10.48550/arXiv.2411.09969
20. L. Taylor, D., et al.: Personalized and adaptive learning. in: innovative learning environments in STEM higher education, pp. 17–34 Springer (2021)
21. Lundberg, S.: SHAP Documentation. https://shap.readthedocs.io/en/latest/
22. Molnar, C.: LIME. In: Interpretable Machine Learning: A Guide for Making Black Box Models Explainable. Christoph Molnar (2025)
23. Nye, B.D., et al.: AutoTutor and family: a review of 17 years of natural language tutoring, vol. 24, pp. 427–469 (2014). https://doi.org/10.1007/s40593-014-0029-5
24. Olney, A.M.: GnuTutor: an open source intelligent tutoring system based on AutoTutor. Presented at the Cognitive and Metacognitive Educational Systems: Papers from the AAAI Fall Symposium (FS-09-02) (2009)
25. Olney, A.M.: GnuTutor: an open source intelligent tutoring system based on AutoTutor
26. OLNEY, A.M.: GnuTutor: an open source intelligent tutoring system (Interactive Event). Presented at the (2009)
27. Raza, S., et al.: The future of learning: building trust and transparency in AI education. J. Manag. Pract. Humanit. Soc. Sci. **8**, 3, 62–74 (2024). https://doi.org/10.33152/jmphss-8.3.6
28. Ribeiro, M.T., et al.: Anchors: high-precision model-agnostic explanations. Presented at the April 25 (2018). https://doi.org/10.1609/aaai.v32i1.11491
29. Seclea: Limitation of Explainable AI. https://seclea.com/resources/seclea-blogs/limitations-of-explainable-ai/

30. Selvaraju, R.R., et al.: Grad-CAM: visual explanations from deep networks via gradient-based localization. https://arxiv.org/abs/1610.02391, (2019). https://doi.org/10.1007/s11263-019-01228-7
31. Wang, D., et al.: Using explainable AI to unravel classroom dialogue analysis: effects of explanations on teachers' trust, technology acceptance and cognitive load. Br. J. Educ. Technol. **55**(6), 2530–2556 (2024). https://doi.org/10.1111/bjet.13466
32. Wilson, K., Nichols, Z.: The knewton platform a general-purpose adaptive learning infrastructure. KNEWTON WHITE Pap (2015)

Analyzing Student Feedback to Assess NoSQL Education

Vanessa Meyer[✉][iD], Lena Wiese[iD], and Ahmed Al-Ghezi[iD]

Computer Science Institute, Goethe University Frankfurt, Frankfurt, Germany
v.meyer@em.uni-frankfurt.de, {lwiese,alghezi}@cs.uni-frankfurt.de

Abstract. We present a learning analytics study on a Master-level practical database course designed to equip students with hands-on experience in the assessment of four database systems for specific use cases. The course is based on a React web application to monitor task performance, including success rates, executability, processing time, and perceived difficulty. Learning analytics conducted across two semesters reveals trends in student success and challenges, such as superior performance in Schema Evolution tasks with PostgreSQL and increased difficulty in Network Analysis tasks across all databases. While Cassandra's lack of join capabilities introduces additional learning complexities, Neo4J demonstrates constantly higher executability and ease of syntax.

Keywords: Digital Learning Tool · NoSQL Databases · Learning Analytics · Blended Learning

1 Introduction

Digital Learning Tools enable student to interactively obtain new insights by following a learning path in a task-based manner. In this paper, we present recent progress made in our Not-only-SQL (NoSQL) Databases learning tool which integrates four database systems in a web frontend; at the same time our system collects statistics and feedback from student to enable Learning Analytics.

Related Work. Only some related systems present a graphical user interface integrating several database system – however, without supporting in-depth learning analytics statistics in the same way as our system does.

First of all, [1] introduces TriQL, a system designed to help students understand and compare three major database models—relational (MySQL), graph (Neo4J), and document-oriented (MongoDB). It provides a graphical query builder that allows users to design queries without prior database programming experience; the presented tool is able to translate them into SQL, Cypher, and MongoDB queries. TriQL's architecture includes an intermediate query generator (using DataLog) and a schema translator.

Moreover, the DBQL tool [3] is meant for teaching relational, document-oriented, and graph-based data models along with their query languages (SQL,

MongoDB Aggregation Pipelines, and Cypher) by offering self-paced, automated exercises. It is based on an adaption of the widely-used Mondial database; this allows direct comparisons between different data models and their query methods on the same dataset. A case study with two student groups (computer science and interdisciplinary majors) demonstrated the effectiveness of this method; interestingly, SQL was rated as the easiest and most relevant query language.

A recent study [2] analyzes error patterns in database query tasks. The study analyzes query submissions from 462 students across SQL, MongoDB, and Neo4j to pinpoint difficulties in relational, document, and graph query languages. The major finding is that aggregation and join operations are the most difficult across all three database models; the authors claim that this confirms prior research on SQL difficulties and extends these to NoSQL systems. The research introduces a concept mapping framework (Projection, Selection, GroupBy, Join) to categorize student errors; the analysis shows MongoDB's higher rate of incorrect result set errors compared to SQL and Neo4j.

Contributions. In [7] Wiese et al. present a teaching design for NoSQL databases. We extend the teaching design with a digital learning tool to collect students' performance data on database tasks and improve the learning experience of NoSQL databases.

In [5] the NoSQLconcepts tool is presented. The architecture of the web application and the technologies used are discussed in detail. An initial usability analysis during one semester is also presented.

The paper [4] focuses on the syntax highlighting of students' query inputs in the NoSQLconcepts tool and the role that syntax highlighting plays in this context.

In contrast to our previous work [4,5], our new contribution is less technical. This paper summarizes the main technologies used and some new components. Our main focus is on the use of the tool in a practical database course and the analysis of the data obtained from the tool.

In addition to usability analysis, our previous work [6] also includes an initial analysis of learning patterns and task difficulty collected by the NoSQLconcepts tool. Data from one semester are considered, and the analysis is based on individual tasks that have not yet been summarized into topics such as Equi-Join. In this paper, we group individual tasks into topics and also add data from an additional semester to our analysis.

Outline. In Sect. 2 we present the Learning Design, which describes a practical course for master students of computer science and master students of business informatics, in which students learn four different NoSQL databases. In addition, the NoSQLconcepts learning tool used in the course and its new functions are described. This is followed by an overview of the tasks and use cases, in which a specific task is discussed as an example. Furthermore, the data collected with the learning tool from the summer semester 2024 and winter semester 2024/2025 is briefly described, which is finally examined in more detail in Sect. 3. Finally, we give a brief conclusion with an outlook for the future.

2 Methodology

2.1 Learning Design

Within our practical course, which is based on the learning concept of Wiese et al. in [7] and is aimed at master students of computer science and business informatics, the basic concepts of NoSQL databases are taught through interactive exercises and hands-on projects. The aim of the course is to provide students with a comprehensive introduction to various database management systems and to familiarize them with the differences between the various databases.

During the assignment phase, students successively work on 10 to 17 tasks in the learning tool, which is described in more detail in Sect. 2.2, which are carried out in the four databases PostgreSQL, Cassandra, Neo4J, and MongoDB. Students are required to independently find suitable queries for the respective task within a specified time. However, there are also tasks that cannot be solved directly with each of the databases (e.g., Join with Cassandra). These limits should be recognized by students, and in such cases, students should give a corresponding explanation as to why a task cannot be solved with a particular database and the given schema.

The practical course is organized as follows. In an initial introductory meeting, students receive an overview of the content and organization of the course. During the semester, weekly online meetings are also held with the course instructor, in which student questions can be addressed, and the current or upcoming exercise sheet with the associated data model is discussed. The aim of these meetings is to enable students to differentiate between the various database systems and understand the underlying data models.

Assignments are another component of the practical course. Students generally have two weeks to complete each database assignment. The assignments are completed independently within the learning tool. There are also two lab assignments (PostgreSQL and Neo4J), which are completed on-site under the supervision of the course instructor. The aim of the assignment phase is for the students to acquire the ability to work practically with various database systems and to be able to independently formulate suitable queries in each of the four query languages for a given task.

Finally, students work alone or in a group on a self-management project. Within the final practical project phase, a web application is developed by each student (or group) in conjunction with a database on a selected topic. At the end of the project phase, the students present their project results. The project phase should enable students to make a well-founded judgement as to which database system is best suited for specific use cases.

2.2 React Web Application

The practical course described in Sect. 2.1 is supported by the NoSQLconcepts learning tool. The learning tool provides students with a standardized environment in which they can complete exercises on four different NoSQL databases.

Figure 1 and 2 show screenshots of a task in an exercise sheet. The exercise sheets also collect data on the time needed for a task, level of difficulty and self-assessment of the correctness of the solution. This data and its analysis will be considered in the further course of this work.

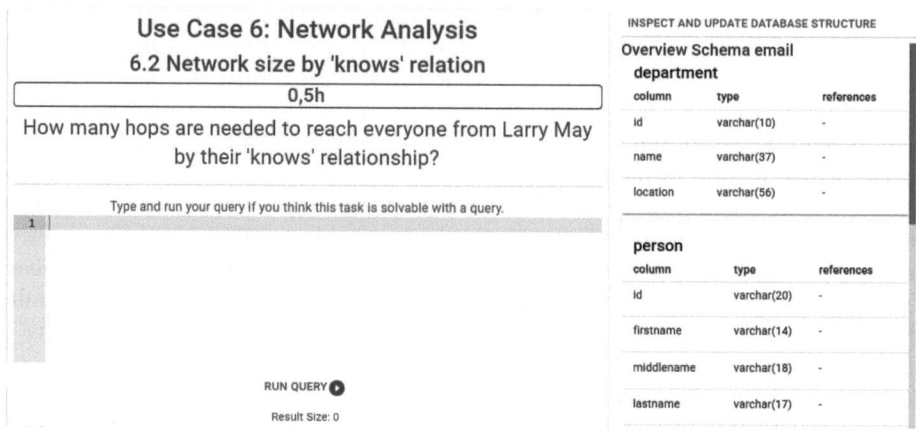

Fig. 1. Screenshot of a task in the exercise sheet of the NoSQLconcepts tool, including the task details, overview of the database schema and an editor for query input.

The learning tool is a React web application that uses the technologies React[1], Node[2], Express[3], together with the databases PostgreSQL, Cassandra, Neo4J and MongoDB as a stack, with React providing a user-friendly interface and Node enabling communication with the respective databases. For further technical details and more information on the architecture of the web application, as well as the existing components and functions of the learning tool, we refer to our previous work [4–6]. The source code of the react application is also available in Github[4].

In this section, we would like to take a closer look at the latest additions to the learning tool. To prevent solutions from being accidentally overwritten when leaving the page if this is not desired, a dialogue window has been added for reconfirming the entries. In this way, users can be sure that those entries that meet their expectations are saved. This change was made in response to student feedback from the winter semester 2024/2025, which showed that solutions were overwritten or not saved.

In addition, participants developed the following components for the learning tool as part of the self-study projects in the winter semester 2024/2025: One of the projects dealt with the display of the query runtime in the interactive exercise

---

[1] https://react.dev/.
[2] http://nodejs.org.
[3] http://expressjs.com.
[4] https://github.com/VaneMeyer/nosqlconcepts.

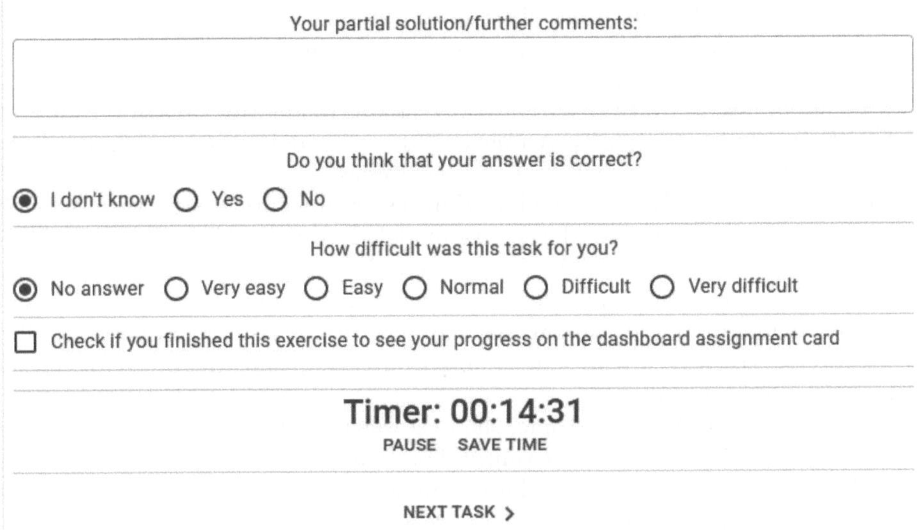

```
                    Your partial solution/further comments:
┌─────────────────────────────────────────────────────────────┐
│                                                             │
└─────────────────────────────────────────────────────────────┘

                    Do you think that your answer is correct?
  ● I don't know    ○ Yes    ○ No

                    How difficult was this task for you?
  ● No answer   ○ Very easy   ○ Easy   ○ Normal   ○ Difficult   ○ Very difficult

  ☐ Check if you finished this exercise to see your progress on the dashboard assignment card

                          Timer: 00:14:31
                          PAUSE   SAVE TIME

                            NEXT TASK >
```

Fig. 2. Screenshot of the second part of the exercise sheet, including a textfield for user comments as well as radio button groups for the self-assessment of correctness/difficulty.

sheet and the function to cancel a query that has been running too long. Another project was concerned with a user profile so that users receive further details on their individual learning progress, such as the task rated by the user as the most difficult/easiest, or the task and assignment with the longest time required. In a third project, a survey component was developed in which the course instructor can create and edit surveys and questions on usability that can be completed by the users. The user responses and statistics are automatically visualized for the instructor with the help of various charts so that no external survey tool needs to be used in future and the automatic creation and visualization of statistics saves the instructor time for evaluating the survey data.

2.3 Use Case and Tasks

We used a subset of the Enron corpus (a large set of email messages) but expanded the original single CSV files with additional connections to model a graph structure as well as allow for foreign key relationships (in the tabular representation). This graph structure was obtained by extraction the information of individual persons as well as their connections by having sent emails to one another as well as adding complementary information on departments for each person.

To showcase and stresstest the advantages of each type of database system, the following task categories ("topics") were developed for the exercises:

- Equi-Join (EJ): In this topic, data from different sources were joined based on matching specific attributes, such as combining a list of employees with their respective departments.
- Missing Values (MV): In this topic, the task was to identify gaps in the dataset for attributes in the emails. A query goal was to determine which attribute contained the highest number of missing values.
- Network Analysis (NA): To explore Enron's organizational structure through email interaction, each employee was modeled as a node in our graph structure, with connections specified by a new "knows" relationships defined by email exchanges. Tasks include measuring graph size via the knows relation, as well as identifying all persons reachable within a 2-hop email network, and identifying employees who sent emails to a certain amount of recipients by counting outgoing edges.
- Range Query (RQ): In this topic, the task was to retrieve data that satisfies a condition involving a range of values. For example, specifying a salary range to find all employees whose salaries fall within that range; or specifying a time range to identify emails sent between specific time intervals.
- Schema Evolution (SE): In this topic, the task was to expand the data structures e.g. by introducing new attributes.
- Theta-Join (TJ): This topic involved combining data based on comparisons like analyzing whether individuals who earn more than their department's average salary send a larger volume of emails compared to those earning less.
- User-Defined Function (UDF): In this topic, the task was to develop a User-Defined Function for generating word counts and incorporating it into the database for analysis.

An Example Problem. Here we show an example task on the topic of network analysis. We also show the solutions from the course instructor for the four assignments and thus the four different databases. We then summarize how the students solved the task, and which errors occurred frequently.

- Problem description: How many hops are needed to reach everyone from Larry May by their "knows" relationship?
- Instructor's solution with SQL:

```
select distinct(person.id)
from email.person as person, email.knows as knows
where knows.pid2=person.id and knows.pid1
in (select distinct(person.id)
from email.person as person, email.knows as knows
where knows.pid2=person.id and knows.pid1
in  (select distinct(person.id)
from email.person as person, email.knows as knows
where knows.pid2=person.id and knows.pid1='may-1'))
```

- Instructor's solution with CQL: The network analysis tasks require all a change in the data model and are impossible to solve given that no joins are available.

– Instructor's solution with Cypher:

```
MATCH (p1:Person {firstname : "Larry"}), ( p2:Person)
WHERE p2.firstname <> "Larry"
WITH p2, length (shortestPath((p1)-[:KNOWS*]-(p2)))
AS myDist
RETURN p2.firstname, p2.lastname, myDist
```

– Instructor's solution with MQL:

```
db.knows.aggregate([
{ $match: {pid1: 'may-1'} },
{ $graphLookup: {
from: 'knows', startWith: '$pid1',
connectFromField: 'pid2', connectToField: 'pid1',
as: 'connections', maxDepth: 2, } },
{ $unwind: "$connections" },
{ $group: { _id: "$connections.pid2" } },
{ $count: "PersonCount" } ])
```

Looking at the students' solutions to this specific task, many problems with recursions occurred with SQL, which led to infinite loops or timeouts. However, there were hardly any syntax errors. Most of the errors were semantic.

For the task with Cassandra, most students correctly recognized that Cassandra does not support this type of task (with join or recursion). Some students suggested alternative options using external tools such as Python. Only a few solutions contain concrete query attempts with Cassandra's CQL, but these are not successful or do not lead to a suitable solution to the task.

Neo4J is very well suited for the given task, which can also be seen in the students' solutions, as almost all solutions are executable and provide correct results, mostly using the shortestPath method.

The solutions to the task with MongoDB show that MongoDB is not ideal for such a task, some suggest alternative options with external tools such as Python or even manual steps. Few solve the task correctly using graphLookup. In the future, we will continue to systematically analyze the students' solutions.

2.4 Learning Analytics Collected Data

In a previous paper [6], we analyzed the data collected using the learning tool for the first time. In contrast to individual tasks, in this paper we want to analyze the data grouped according to the respective topic (e.g. Equi Join, Theta Join). In addition, in this paper we combine the data from two semesters (a total of 23 participants) so that we analyze a larger data set than in the previous paper.

To ensure an understanding of the data collected with the tool in this paper, we will briefly discuss the features of the data set used for the learning analysis. The analysis is an exploratory analysis in which we look at the mean values across all participants, grouped by database assignment and topic, for the features

Success Rate, Executability Rate, Processing Time, which is the time needed to solve a task, and *Difficulty Rates*. The success rate is based on the student's self-assessment. For each task they complete in the learning tool, they provide an assessment of whether they have solved the given task correctly. Students can choose between three answer options: *I don't know, Yes,* or *No,* whereby only the answer Yes is counted as a success in the analysis. To calculate an average success rate, the categorical values are assigned to binary values so that the answer option Yes is represented by the value 1 and the other two options are given the value 0 for no success.

We proceed in the same way to calculate the executability rate. If the queries entered by users can be executed when running it in the learning tool, information on the executability of the query for the respective task is saved in binary form (1 if the query is executable, otherwise 0). The learning tool, or more specifically the digital exercise sheet, has an integrated timer component that automatically measures the time as soon as a task is started. The time is saved in seconds. These are then converted into minutes for a clear overview.

Students can also specify the perceived difficulty of a task in the learning tool. The options are *very easy, easy, normal, difficult* and *very difficult.* These are also transformed into numerical values, with very easy being assigned the lowest value of 1 and very difficult the highest value of 5.

3 Results and Discussion

In this section, we will discuss the results of the evaluation of the learning analytics data collected in our learning tool. We take a look at the individual topics of the exercises in terms of processing time, level of difficulty, self-assessment of success and executable queries, to find out with which topics students had more difficulties and to use this knowledge to adapt exercises and tutorials accordingly.

3.1 Self-assessed Success Rates

Firstly, we take a closer look at the success rates for each database and topic. As can be seen in Fig. 3, not every topic is represented in every database assignment. This is due to the fact that two databases are intensely covered in the on-site lab sessions as well.

Looking at the PostgreSQL exercises (red), the success rates are highest for Schema Evolution, followed by the Equi Join exercises. The Network Analysis exercises, on the other hand, have the lowest success rate. This indicates that students feel comfortable with the syntax of the Schema Evolution tasks, which involve adding, deleting, or editing tables, as the general syntax of such queries does not usually have an overly complicated structure. The given Equi Join tasks also serve as an introduction to the respective assignment and expect a simple query, which is reflected in the success rate. In the Network Analysis tasks, somewhat more complex queries are expected to solve the exercise task,

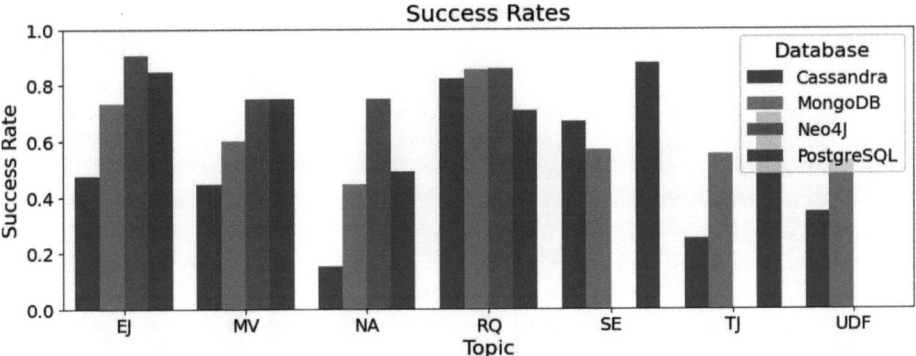

Fig. 3. Success rates, based on students self-assessment, for each database assignment and exercise topic. The topics are: Equi Join (EJ), Missing Values (MV), Network Analysis (NA), Range Queries (RQ), Schema Evolution (SE), Theta Join (TJ) and User Defined Functions (UDF). (Color figure online)

so that the relatively low success rate here also corresponds to the teacher's expectations.

If we look at the exercises in the Cassandra database (blue), we can already see differences to the results of the PostgreSQL database. The Schema Evolution tasks no longer have as high a success rate as before with PostgreSQL and the success rate for Equi Join tasks is relatively low. The success rate for Network Analysis is also much lower than for the PostgreSQL tasks. However, with Cassandra, the success rate for Range Queries is the highest and even slightly higher than with PostgreSQL. These results suggest that the fact that Cassandra does not support join operations, influences some of the students regarding their self-assessed success, as the students seem to be unsure how they can solve the join tasks without a join. In this case, teachers expect students to find out for themselves that the specified join tasks cannot be solved in Cassandra.

Neo4J (green) has a relatively high success rate in all topics, indicating that the Cypher query syntax is easy to learn for students.

With MongoDB (orange), Range Query tasks have the highest success rate, followed by Equi Join; these tasks do not require complex queries to be solved, which is consistent with the success rates. In addition, the rates for Schema Evolution tasks are even lower than for PostgreSQL and Cassandra, and the Network Analysis tasks also have a relatively low success rate compared to the other MongoDB topics. We have already mentioned in Sect. 2.3 that MongoDB does not seem to be ideal for Network Analysis tasks.

3.2 Rates for Executable Database Queries

Figure 4 shows the average number of executable database queries across all students and for each database assignment and topic.

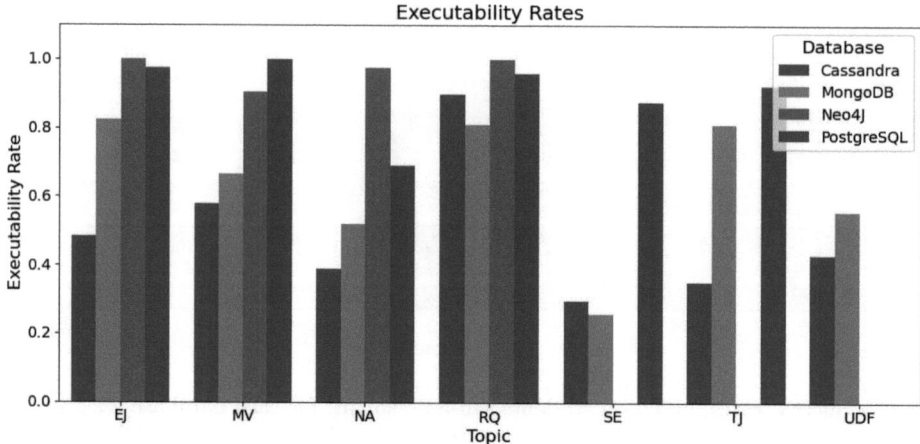

Fig. 4. Rates of executable database queries for each database assignment and exercise topic.

In the PostgreSQL assignment (red), we have an average rate of 1, that is 100%, for the Missing Values topic. This means that all students have found an executable query for these tasks. The Equi Join, Range Queries and Theta Join also achieved a very high rate. Although Schema Evolution tasks achieved the highest success rate for the PostgreSQL assignment as described above, the rate in terms of executability is slightly lower than for Equi Join, Range Queries and Theta Join. This could be due to errors during execution in the web application, or to an incorrect self-assessment of success. In any case, the instructor can focus on the submissions on this topic to uncover potential problems. As for the rate for network analysis, the relatively low value is in line with the success rate.

For Cassandra assignments (blue), all rates for executability are very low compared to PostgreSQL assignments, except for the Range Queries topic. This also reflects the previously considered success rates and the non-support of join operations in Cassandra.

Compared to the other database assignments, Neo4J (green) has very high rates, with Equi Join and Range Queries achieving a 100% rate for the executability of database queries. This is also in line with the results of the success rates and emphasizes simplicity of Cypher syntax.

For the MongoDB assignment (orange), the topics Equi Join, Theta Join and Range Queries have the highest values, while Schema Evolution and Network Analysis have the lowest. This is also largely in line with the success rates, although the rate for Schema Evolution is very low compared to the success rate. This could be related to the fact that the database driver mongoose used in the web application does not have as much flexibility for executing queries as the other database drivers, especially in relation to schema evolution tasks where collections are to be added or removed. Improvements can be made here in future and, if necessary, a different database driver can be selected to ensure more flexibility in the execution of database queries entered.

3.3 Needed Time to Solve a Task

In addition to the success rate and executability of the database queries, the time required by the students to solve a task was also automatically recorded with the learning tool. The average time required for individual topics of the database assignments is shown in Fig. 5. On average, students need less than 60 min for most assignment tasks. However, the time required for Theta Join tasks in the PostgreSQL assignment stands out in the figure, which represents an unexpectedly high number of minutes at first sight. If we consider the maximum time of 2 h specified by the course instructor for the Theta Join task, the average time required by the students is still significantly less than the specified time. In addition, the specified maximum time for the Theta Join task is set higher than for other tasks, so that this is still in line with the instructor's expectations. In addition to the Theta Join tasks, the Network Analysis and User Defined Function tasks also have slightly higher times than Equi Join, Missing Values, Range Queries, and Schema Evolution, which also reflects the specified maximum times (Network Analysis tasks have a maximum time around 1 to 2 h, while Equi Join, Missing Values, Range Queries have a maximum time of around 0.5 h). Students usually need even less time, so that the instructor could consider adapting the tasks.

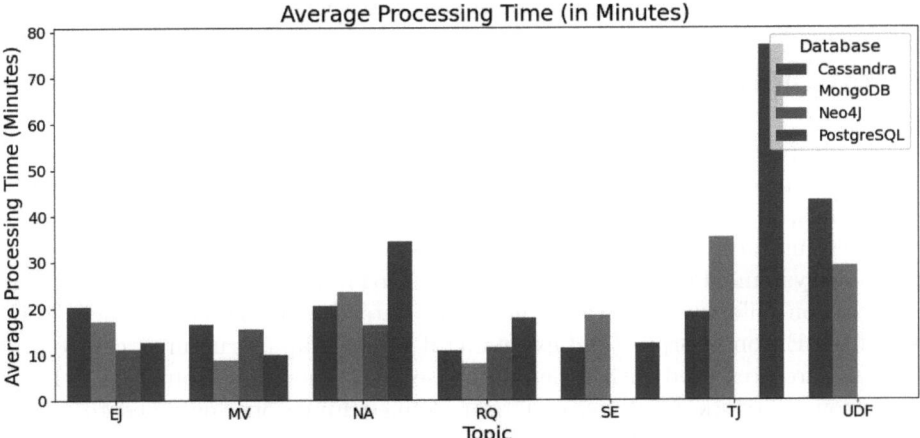

Fig. 5. Average time needed for tasks for each database assignment and exercise topic.

3.4 Difficulty Levels

Regarding the average perceived difficulty, Fig. 6 shows that Network Analysis, Theta Join, and User Defined Functions areas have a slightly higher perceived difficulty level overall than the other topics. Neo4J also has the lowest difficulty level in the categories Equi Join, Missing Values, Network Analysis, and Range Queries compared to the other database assignments, which is consistent with our previous knowledge of the simplicity of Cypher syntax.

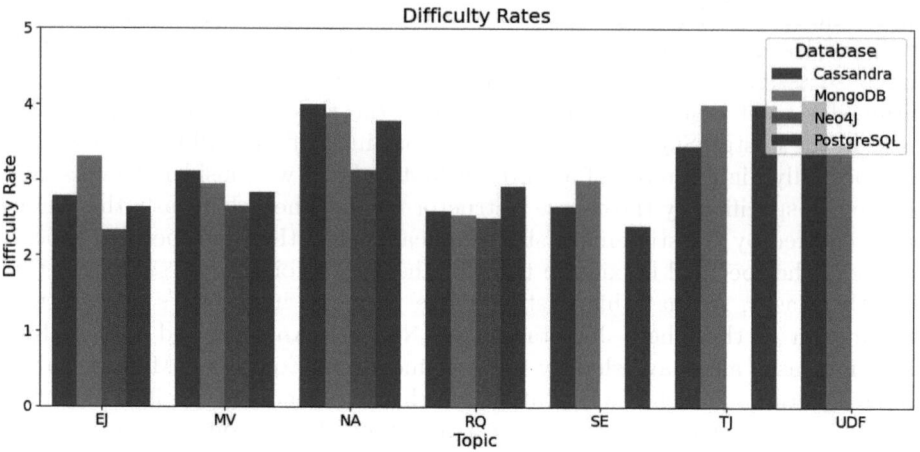

Fig. 6. Average difficulty-level, based on students self-assessment, for each database assignment and exercise topic.

4 Conclusion

In this study we presented a comprehensive assessment of student performance in a practical course aimed at master's students in computer science and business informatics; in order to introduce NoSQL database concepts through hands-on exercises, interactive tasks, and a self-management project, the course covers four database systems: PostgreSQL, Cassandra, Neo4J, and MongoDB. The learning focus lies on differentiating their underlying data models and capabilities. Using data from Enron's corpus (and extending it into a graph structure), course participants are presented with varied database tasks, including Equi-Join, Schema Evolution, Network Analysis, and User-Defined Functions, designed to stress-test the advantages and limitations of each system. In future work we aim to get more insights from the analysis of error patterns in the student submission. This will help us in refining the course tasks and in enhancing the overall learning outcomes.

Acknowledgments. This paper was supported by Goethe University Frankfurt's Digital Teaching and Learning Lab (DigiTeLL) in the project NoSQLConcepts funded by Stiftung Innovation in der Hochschullehre.

References

1. Alawini, A., Rao, P., Zhou, L., Kang, L., Ho, P.C.: Teaching Data Models with TriQL. In: Proceedings of the 1st International Workshop on Data Systems Education, pp. 16–21 (2022)
2. Alkhabaz, R., Li, Z., Yang, S., Alawini, A.: Student's learning challenges with relational, document, and graph query languages. In: Proceedings of the 2nd International Workshop on Data Systems Education: Bridging Education Practice with Education Research, pp. 30–36 (2023)
3. Ehlers, J.: Teaching multiple data models and query languages. In: Proceedings of the 2024 on Innovation and Technology in Computer Science Education, vol. 1, pp. 234–240 (2024)
4. Meyer, V., Palakat, J.J., Wiese, L.: Integrating syntax highlighting into a digital nosql database teaching tool. In: Lernen, Wissen, Daten, Analysen (LWDA), pp. 1–13 (2024)
5. Meyer, V., Wiese, L., Al-Ghezi, A.: A unified teaching platform for (No) SQL databases. In: 26th International Conference on Enterprise Information Systems (ICEIS), vol. 1, pp. 149–160 (2024)
6. Meyer, V., Wiese, L., Al-Ghezi, A.: Analyzing learning patterns, task difficulty and usability in a digital database learning tool. In: 2024 21st International Conference on Information Technology Based Higher Education and Training (ITHET), pp. 1–10. IEEE (2024)
7. Wiese, L., Benabbas, A., Elmamooz, G., Nicklas, D.: One DB Does Not Fit It All: Teaching the Differences in Advanced Database Systems (2021)

Data Mining

Data Mining

Data Mining for Language Superfamilies Using Congruent Sound Groups

Peter Z. Revesz[1,2](✉) and Mohanendra Siddha[1]

[1] School of Computing, University of Nebraska-Lincoln, Lincoln, NE 68508, USA
[2] Department of Classics and Religious Studies, University of Nebraska-Lincoln, Lincoln, NE 68508, USA
prevesz1@unl.edu

Abstract. There have been several attempts in recent years to prove that some well-known language families can be grouped together into a language superfamily. This paper presents a data mining method to search for a language superfamily. The data mining method is based on the consideration of regular sound changes in various languages. Congruent Sound Groups are derived from the commonly observed regular sound changes. To demonstrate the feasibility of this method, we collected a set of words related to the four basic elements of air, earth, fire and water from seven languages: Hindi, Japanese, Korean, Russian, Sanskrit, Tamil and Telugu. These seven languages are classified into four different language families: Dravidian, Indo-European, Japonic, and Koreanic. The congruent sound group-based analysis enabled the identification of seven cognate groups of words that involve different language families. This suggests that these four different language families originate from a single protolanguage that was likely spoken in Asia more than 10,000 years ago.

Keywords: Data mining · Language family · Congruent sound group · Regular sound change

1 Introduction

Traditional linguistics has classified many languages into language families. Many methods used by traditional linguistics for building language family trees can be automated as shown by Bouckaert et al. [3] and Ciobanu and Dinu [5] for the Indo-European language family, Honkola et al. [12] for the Uralic language family, Kolipakam et al. [16] for the Dravidian language family, and Mutabazi and Revesz [20] for the Bantu language family. Traditional linguistics is also interested in analysing languages for the degree of vowel harmony that they have, and that analysis can also be automated [39] and also applied to the analysis of the underlying languages of ancient scripts [31].

However, computer science can aid not only in automating the traditional linguistic methods for finding traditional language families and language family trees but also in developing novel algorithms for finding language superfamilies, which bring together language families that were previously considered separate.

Miyagawa et al. [19] have shown recently that language capacity was present in humans at least 135 thousand years ago, which is before the spread of modern humans from a common homeland in Eastern Africa. That naturally suggests that many of the currently known language families have a common root. However, linguistically identifying superfamilies is a difficult task because the traditional methods of grouping languages into a language family no longer seem to be applicable. Vajda and Fortescue [38] found some connections between Siberian and Native American Indian languages using traditional methods, but the number of cognate words that they found still seems to be too few for a general acceptance of their theories by fellow linguists. Hence, it is a welcome new research direction to develop novel methods that can show more convincing connections among language families.

Gell-Mann and Ruhlen [10] considered the issue of how the subject, verb and object word order in languages could have derived from a common root. They showed that the word order could have changed such that the changes largely align with traditional linguistic classifications into language families. However, word order can vary even within a single language family and is also known to have changed in the history of particular languages. Hence, word order is not the best way to approach the issue of identifying language superfamilies.

Pagel et al. [22] considered seven Eurasian language families and suggested twenty-one words from a hypothetical common protolanguage from which those seven Eurasian language families may originate. While Pagel et al.'s results are interesting, their method is just an exhaustive search instead of a more efficient method.

Yang et al. [41] developed a velocity field estimation method that analyzed about 700 languages and classified them into four groups. However, these groups are not superfamilies but groups that are loosely associated with four agricultural areas where the languages could borrow words from each other. In contrast, language superfamilies are supposed to be based on common origins rather than word borrowings.

Schaefer and Revesz [36] proposed a fuzzy morphism matching algorithm that considers a graph representation of the phonetic structures of various languages. They use a type of fuzzy graph isomorphism matching algorithm to find distant connections between languages.

In this paper, we propose a new data mining method that can efficiently search for groups of cognate words. This method is based on the idea of using congruent sound groups to efficiently identify similar words. We also focus our search for the words that are related to the four basic elements of air, earth, fire and water. Of course, modern science recognizes more than these four elements, but ancient people from Europe to Asia recognized only these four basic elements since a very early time. Presumably, the recognition of these four basic elements influenced the development of the vocabulary of a common superfamily protolanguage in Eurasia. That means, that the various words related to water in the current descendant languages could be related to each other even if they do not have the same meaning. For example, a lake and a creek are different objects, as is obvious from having two different names for these objects in English. Other modern languages also distinguish these two objects.

However, it is possible that the superfamily protolanguage contained only one word for water of all kinds. Then the protowords' meaning narrowed to mean 'lake' in one language family, while its meaning narrowed to mean 'creek' in another language family. Moreover, it could have preserved its more generic meaning of 'water' in a third language family. Hence, the traditional approach of treating the words meaning 'lake', 'creek' and 'water' in the three language families as unrelated words even when they sound similar is unwarranted. Instead, we allow a more flexible approach, where the meanings are interchangeable as long as they all somehow relate to the element of water. We treat the other groups of words related to air, earth, and fire in a similarly flexible way.

The rest of this paper is organized as follows. Section 2 gives an overview of phonetic notations of the *International Phonetic Alphabet* [13] and basic concepts such as regular sound changes and congruent sound groups. It also describes our data sources of words from seven different languages classified as four different language families. Section 3 describes the experimental method and results. Section 4 discusses the results. Finally, Sect. 5 gives some conclusions and directions for further work.

2 Phonetic Notations and Data Sources

This section first reviews the *International Phonetic Alphabet* [13], which is used for the phonetic notation of the words studied in this paper. Then it gives an overview of the concepts of *regular sound changes* and *congruent sound groups*. Finally, it describes the selection criteria and the sources of the words that we studied.

2.1 The International Phonetic Alphabet

The words used in this study are recorded using many different alphabets that may not reflect the current pronunciation of the words well. Moreover, these alphabets are very different from each other. Hence, linguists frequently use a standard notation called the *International Phonetic Alphabet* or IPA for describing the pronunciation of words.

Many of the IPA signs for the various phonemes are the same as the English alphabet letter that usually expresses the same phoneme. In these cases, the only distinction between the IPA phonetic notation and the English alphabet letter is that the phonetic notations are placed in brackets such as /b/, which expresses the phoneme at the beginning of the word 'butter.' However, there are some differences from the usual English pronunciation and some special IPA signs such as the following:

ð - The sound in the middle of *father*.
j - The second phoneme in the word *fjord*.
ŋ - The sound at the end of *sing*.
ʃ - The sound at the beginning of *sheep*.
θ - The sound at the beginning of *thin*.
ʒ - The sound in the middle of *vision*.

Since the typing of these special notations is often difficult, the second special sign is often replaced by š, and the fourth special sign is replaced by ž. We will also use these sometimes in the rest of this paper.

The IPA classifies the various phonemes according to several criteria that include the manner and place of pronunciation. The IPA also gives a useful table that classifies the phonemes based on the manner and the place of pronunciation. Table 1 shows a simplified version of the IPA table.

Table 1. A simplied International Phonetic Alphabet table.

	Bilabial	Labiodental	Dental	Alveolar	Postalveolar	Palatal	Velar	Glottal
Plosive	p, b			t, d			k, g	
Nasal	m			n			ŋ	
Trill				r				
Fricative		f, v	θ, ð	s, z	ʃ, ʒ			h
Approximant						j		
Lateral appr.				l				

Palatized phonemes are described by adding a j. Here are a few examples:

dj - The sound at the beginning of *dew*.
lj - The sound at the beginning of *lute*.
nj - The sound at the beginning of *new*.
sj - The sound in the middle of *consume*.
tj - The sound at the beginning of *tune*.
zj - The sound in the middle of *resume*.

Another frequently used compound notation is the following:

tʃ - The sound at the beginning of *chase*.

The manner of pronunciation of the consonant phonemes can be *plosive*, which means that there is a complete closure and a sudden release during the production of the sound, *nasal*, when the air travels through both the nasal cavity and the mouth, a *trill* when the tongue vibrates against the alveolar ridge, *fricative*, when the airflow is impeded so much that is creates a hissing-like sound, an *approximant*, when the airflow is only little impeded and no hissing sound is created, or *lateral*, when it is like a plosive with some space left on the side for the airflow. Instead of being a trill /r/, the English pronunciation of words with the r letter is usually a voiced postalveolar approximant, which is indicated by an upside-down r letter.

The place of pronunciation of the consonant phonemes can be *bilabial*, when produced by the lips alone, *labiodental*, when a combination of the lips and the teeth are used, *dental*, when the tongue touches the teeth, *alveolar*, when the tongue is behind the teeth, *postalveolar*, when the tongue is further back from the teeth, *palatal*, when the tongue touches the top of the palate, *velar*, when the tongue is at the back of the mouth, or *glottal*, when the throat is compressed.

The traditional analysis of phonemes from all the human languages tried to place every observed phoneme into a neat category regarding manner and place of pronunciation. However, there are many borderline cases that are hard to classify. Hence, it may be better to think of a fuzzy representation of the phonemes on several scales, for example the degree of being completely at the front or classic bilabial would be 0 and being completely produced in the throat would be 1, and the rest could be given a value on the range between 0 and 1. This idea and its advantages for speech recognition are explored in [34].

The IPA also gives a list of notations for various vowels. We skip a review of the vowel notations because the most important regular sound changes that are used to show the relationship among languages involve only consonants.

2.2 Regular Sound Changes

The usual method to prove that languages have a common origin is to collect words from those languages that are likely to be cognates, that is, words that are likely to have a common origin because their meanings and pronunciations are the same or similar. The case of being only similar rather than identical is due to the fact that the languages developed differently after the common protolanguage separated into different branches. This development, however, is not haphazard but regular in the sense that the same phoneme in the same position undergoes the same change in all cognate words within the same language.

For example, Latin *pater*, *ped*, and *porro* are cognates with English *father*, *foot* and *far*, respectively. In these words the word initial /p/ in Latin always corresponds to the word initial /f/ in English. We call this phenomenon a *regular sound change*. Regular sound changes can be detected using association rule mining methods [29], which were applied to detect regular sound changes between Uralic languages [1, 26, 32] and the Pre-Greek substrate of ancient Greek ([30] and Tables 13-14 in [35]), which has a mostly Minoan origin, and Sumerian [33] based on the etymological dictionaries of Beekes [2] and Parpola [23], respectively.

Regular sound changes can go in either direction because a phoneme can weaken or strengthen over time in a spoken language, that is, it can undergo either lenition or fortition. There are cases when we need to investigate a set of words for potential regular sound changes, but we do not know the direction of sound changes, for example, whether there are lenitions or fortitions. Moreover, these two processes as well as a number of other processes could reverse themselves during the development of a language. Still, we tend to see some similar patterns in many language families.

For example, let us consider Finnish and Hungarian, which are the two most frequently spoken Uralic languages. Finnish *pää* 'head' and *pilvi* 'clound' are cognates with Hungarian *fej* 'head' and *felhő* 'cloud', respectively. Clearly, the word initial /p/ in Finnish changes to Hungarian /f/ similarly to how the Latin word initial /p/ changed to English word initial /f/.

Depending on the commonality of the change between /p/ and /f/ in several different language families, we may consider these two phonemes to form a group.

2.3 Congruent Sound Groups

Protowords can change over time by the replacement of some group of phonemes by another group of phonemes. We saw in the previous section that some of these phonetic changes are more frequent than others in the various linguistic families.

In general, those phonemes which are very similar can change more easily to each other than those which are less similar. A pair of consonants can be similar because they are voiced and unvoiced pairs such as the consonants /b/ and /p/. A pair of consonants can be also similar because they have the same location of pronunciation, such as /b/ and /m/, which are both bilabial, or the same manner of pronunciation such as /m/ and /n/, which are both nasal.

Revesz [28] called a group of consonants that can easily change among themselves during the development of languages a *congruent sound group (CSG)*. In particular, Revesz [28] defined the first six congruent sound groups below to which we added the seventh:

1. /b/, /p/, /m/ and /n/. In this group, the /b/ and /p/ form a voiced-voiceless pair of consonants. They are bilabial, plosive phonemes and can change into the bilabial, nasal phoneme /m/. The /m/ can change into the nasal /n/.

2. /d/, /t/ and /θ/. The /d/ and /t/ form a voiced-voiceless pair of consonants. The /θ/ is an aspirated version of /t/.

3. /f/, /v/ and /w/. The /f/ and /v/ form a voiced-voiceless pair of consonants. The /v/ and /w/ can also be exchanged.

4. /g/, /k/ and /h/. The /g/ and /k/ form a voiced-voiceless pair of consonants. The /k/ can change to /h/ in many language families. For example, /k/ in Proto-Uralic regularly changes to /h/ in Hungarian when /k/ is at the beginning of words and is followed by a back vowel.

5. /l/ and /r/. These phonemes are both alveolar liquid phonemes. These frequently change to each other, and some writing systems, such as the Linear B script, which used to write Mycenaean Greek during the Bronze Age, did not distinguish between them [40].

6. /s/, /z/, /š/ and /ž/. The /s/ and /z/ form a voiced-voiceless pair of consonants. The /š/ and /ž/ also form a voiced-voiceless pair of consonants. All of these are fricative phonemes and can easily change into each other.

7. /j/. This is a semivowel that was not used in Revesz [28] but occurs in some of the words that we considered.

We extended the above CSGs by adding retroflex forms of /d/ and /t/ to CSG 2, the phoneme /ŋ/ to CSG 4, and several variations of the liquid /r/ phoneme to CSG 5. These extensions were necessary because some of the languages that we studied contain these phonemes.

The above set of congruent sound groups does not mean that a phoneme in one group can never change to a phoneme in another group. As we saw in the previous section, the word initial /p/ can change to /f/ within several language families. However, any grouping needs to make some distinction between frequent sound changes, rare sound changes and impossible sound changes. Pellard et al. [25] describe an interesting regular sound change between Greek word-initial /b/ and English /k/ including the examples of Greek *boûs* 'cow' and Greek *baínō* 'come.' While the /b/ and /k/ sound change is clearly regular, it is a rare sound change. Hence, we did not merge CSG 1, which includes /b/, with CSG 4, which includes /k/.

Similarly, the /p/ to /f/ regular sound change was also considered a relatively rare sound change despite the two examples mentioned in Sect. 2.2. Hence, it did not warrant merging CSG 1, which includes /p/, with CSG 3, which includes /f/.

In the rest of this paper, we consider a pair of words from two different languages *cognates* if they have a similar meaning, and they can be transformed from one to the other using exchanges within the above congruent sound groups. We can also derive a phonetic similarity metric based on how many exchanges are needed to transform one word into the other. Kessler [15] and Ladefoged [17] also discuss phonetic similarities and present some alternative phonetic similarity metrics.

We do not review vowel changes in this section because they change more frequently than the consonants.

2.4 Identifying the Basic Vocabulary of Language Superfamilies

We selected seven linguistically diverse languages: Hindi, Japanese, Korean, Russian, Sanskrit, Tamil and Telugu. These represented several different language families. Hindi and Sanskrit belong to the Indo-Iranian branch and Russian belongs to the Slavic branch of the Indo-European language family. Japanese belongs to the Japonic language family. Korean belongs to the Koreanic language family. Tamil and Telugu belong to the Dravidian language family. All of these four language families originate in Asia, including the Indo-European language family [11]. Hence, it is possible that they derive from a common superfamily that existed over ten thousand years ago.

We collected some truly basic vocabulary from the seven languages that may go back to a superfamily language stage. This basic vocabulary consisted of words that were related to the four basic elements of air, earth, fire, and water. That is, we search for the native words from the seven languages that are related to these four basic elements.

- Air. This category includes words like air, breath, and wind.
- Earth. This category includes words like mountain, rock, sand, and soil.
- Fire. This category includes words like ash, fire, flame, and smoke.
- Water. This category includes words like creek, lake, rain, river, and sea.

Our main hypothesis is that the protolanguage of a superfamily of languages had much fewer words than the protolanguages of the language families within that superfamily. There could have been words for basic concepts like 'water' and 'mountain' and 'plain' and 'sky' at the superfamily level, but no separate words for 'spring', 'lake' and 'rain.' Instead, the concept of 'spring' would be described by the compound word 'mountainwater', 'lake' would be referred to as 'plainwater', and 'rain' would be called 'skywater.'

Furthermore, suppose that such a superfamily separated into three branches, which originated let's say the language families M, P and S. The people of language family M may have moved to a mountainous environment, where the primary source of water was various springs. They always associated the original word for 'mountainwater' with 'spring' and no longer used 'plainwater' and 'skywater' as words. Recognizing that mountainwater is a compound word, they just could use 'water' as meaning 'spring.'

Simultaneously, the people of language family P settled in a great plain. They associated the original word 'plainwater' with 'lake'. Then they also recognized that it is a compound word, and then they simply referred to a 'lake' with the original word for 'water.'

Finally, the people of language family S may have moved to an area where there were neither major mountains nor major lakes but had plenty of rainfall. Their language also may have developed so that the superfamily word for 'skywater' and then 'water' started to mean 'rain' in their vocabulary. Eventually, all of the three language families M, P, and S developed a large vocabulary for many types of water.

Traditional linguists follow Swadesh [37], who considered the concepts of 'water', 'rain', and 'lake' to be different *universal concepts*. Hence, in this situation, they would find nothing strange about the fact that the word for 'spring' in the M language family, the word for 'lake' in the P language family, and the word for 'rain' in the S language family were similar to each other. However, what if the set of things that can be considered 'universal concepts' is only relative to the language development stage that is considered? Moreover, what if at the superfamily-level the 'universal concepts' consisted of only a shorter list that included only a general notion of 'water' but not the specific concepts of 'rain', 'lake' and 'spring'? Hence, these three language-family-level 'universal concepts' merge into a single language-superfamily-level 'universal concept' called 'water'?

Hence, identifying language superfamilies requires checking a larger set of possible cross-associations than traditional linguists usually check, but that is easily doable by computational methods.

Revesz [28] already collected some basic vocabulary related to mountains (which is within the earth category) and rivers (which is within the water category). Table 2 shows three groups of words related to mountains collected in [28]. The first consonant is either /p/ or /b/ in all three groups. The second consonant is either /d/ or /t/ in the first group, is /l/ in the second group, and is /r/ in the third group. This suggests that these words are cognates even though they come from five different language families. Furthermore, these suggest that *pata 'mountain, rock' may have been a word in the protolanguage of the superfamily that later separated into the five language families seen in Table 2. The /l/ may have arisen through the sound changes /t/ > /d/ > /l/, or it may be a consonant that was added later for ease of pronunciation. An /l/ > /r/ development is also possible. Of course, it is possible that some of the five language families borrowed these words rather than inherited them from a common superfamily protolanguage. In particular, the Greek words are Pre-Greek [2], which suggests that they were borrowed from the Minoan language, which is a Uralic language [27].

Table 2. Apparently cognate 'mountain, rock'-related words that may derive from a single word in the protolanguage of a superfamily of languages that includes the Afro-Asiatic, Austronesian, Indo-European, Niger-Congo, and Uralic language families.

Word	Language	Language Family	English Meaning	Source
apata	Yoruba	Niger-Congo	rock	[9]
bato	Filipino	Austronesian	rock	[8]
bàd	Sumerian	Uralic	wall	[23]
buda	Hungarian	Uralic	sharp picket, mountain name	[6]
petra	Greek	Indo-European	rock, rocky mountain range	[2]
blat	Maltese	Afro-Asiatic	rocky	[18]
pāl	Mansi	Uralic	pole	[26]
pella	Greek	Indo-European	stone	[2]
pēlp	Mansi	Uralic	pointed	[14]
brdo	Croatian	Indo-European	hill	[4]
būrti	Somali	Afro-Asiatic	rock	[7]

Our work extends the set of basic vocabulary that is considered for analysis. Table 3 shows a subset of the words that we collected from a larger set of over two hundred words. The words in the table were the ones that we found to be cognates in our analysis. All the words are given using the IPA notation for an easier comparison.

Table 3. Sample words (described using the IPA notation) related to the elements of air, earth, fire and water.

Word	Language	Language Family	English Meaning	Category
vaːju:	Sanskrit	Indo-European	air, wind	Air
vaːjuvu:	Telugu	Dravidian	air, wind	Air
kuːki	Japanese	Japonic	air, atmosphere	Air
goŋgi	Korean	Koreanic	air, atmosphere	Air
agnih	Sanskrit	Indo-European	fire	Fire
agni	Telugu	Dravidian	fire	Fire
ogón	Russian	Indo-European	fire	Fire
hono:	Japanese	Japonic	flame	Fire
nadi	Sanskrit	Indo-European	river	Water
nadi	Telugu	Dravidian	river	Water
moː.re:	Russian	Indo-European	sea	Water
mul	Korean	Koreanic	water	Water
niː.lu	Telugu	Dravidian	water	Water
səmudrə	Sanskrit	Indo-European	ocean, sea	Water
samidərə	Japanese	Japonic	early summer rain	Water
gəŋə	Sanskrit	Indo-European	Ganges river	Water
gəŋ	Korean	Koreanic	river	Water

3 Experimental Method and Results

3.1 Method

The lexical items collected from Hindi, Japanese, Korean, Russian, Sanskrit, Tamil, and Telugu were processed as follows:

- Extracting core vocabulary from Swadesh lists [37] and other dictionaries.
- Removing affixes, thereby reducing words to their root forms.
- Categorizing the words as related to one of the four basic elements.

Following the method in Revesz [28], we organize the various words related to each of the four basic elements or categories into four separate tables. Each table has a row and a column for each of the seven CSGs. A word is placed into the ith row if its first consonant belongs to the ith CSG, and into the jth column if its second consonant belongs to the jth CSG. We use a superscript to indicate to which language each word belongs. The legend of the superscripts is: Kor = Korean, Jap = Japanese, Tam = Tamil, Tel = Telugu, Hin = Hindi, San = Sanskrit, Rus = Russian.

Table 4. Cognate groups of words according to the congruent sound groups method.

	b, p, m, n	d, t, θ	f, v, w	g, k, h, ŋ	l, r	s, z, š, ž	j
b, p, m, n		*nadi nadi[San] nadi[Tel]			*more moːre:[Rus] mul[Kor] nilu[Tel]		
d, t, θ							
f, v, w							*vaju vaːju:[San] vaːjuvu:[Tel]
g, k, h, ŋ	*agne agnih[San] agni[Tel] ogón[Rus] honoː[Jap]			*gaŋ gəŋə[San] gəŋ[Kor] *kuŋ kuːki[Jap] goŋgi[Kor]			
l, r							
s, z, š, ž	*samudra səmudrə[San] samidərə[Jap]						
j							

3.2 Results

Table 4 summarizes all the seven groups of cognate words that were found using this method. The groups of cognate words come from different language families. For each group of cognate words, we suggest a hypothetical protoword that is indicated by a * sign at the beginning. The exact form of these hypothetical protowords is debatable, but they do not play a major role in the rest of our analysis. Table 5 shows the protowords of the cognate groups that we found.

The Dravidian and the Indo-European language families are connected by three hypothetical cognate words according to Table 5. The Indo-European language family is connected by two hypothetical cognate words to the Japonic and the Korean language families. Dravidian is connected by one hypothetical cognate word to Japonic and Koreanic. Finally, Japonic and Koreanic are also connected by one hypothetical cognate word. Of course, there is always some possibility that the hypothetical cognate words are caused by some word borrowings. However, such ancient and frequently used words are unlikely to have been borrowed by languages according to Pagel et al. [22]. More words would be needed to be able to securely distinguish between cognate words versus word borrowings and to be able to draw a phylogenetic tree of the entire language superfamily.

4 Discussion

Linguists like to talk about the great linguistic diversity of the world with over five thousand distinct languages [21]. While traditional linguists have classified many of these languages into language families, there is a skepticism among most linguists that humanity once spoke a single world protolanguage. Hence, the linguistic state-of-the-art is in contrast to ancient Indian philosophical teachings that all of humanity is one [24].

Table 5. The hypothetical protowords for the seven cognate groups.

	Dravidian	Indo-European	Japonic	Koreanic
Dravidian		*agne *nadi *vaju	*agne	*more
Indo-European	*agne *nadi *vaju		*agne *samudra	*gaŋ *more
Japonic	*agne	*agne *samudra		*kuŋ
Koreanic	*more	*gaŋ *more	*kuŋ	

The surprising finding that four different language families that were previously considered unrelated are actually related and likely derive from a common protolanguage seems to be an intermediate result. While our results are far from being able to connect all the language families of the world, the results suggests that some superfamily protolanguage existed once in Asia because all four language families that we considered (Dravidian, Indo-European, Japonic, and Koreanic), are thought to originate in Asia.

5 Conclusion

The words for air, earth, fire and water are likely some of the oldest words in any human language. Other old categories of words can be found for example, human beings, food, animals, plants, stars, etc. These categories of words also need to be investigated to find more cognate words. The set of languages considered can also be extended beyond Hindi, Japanese, Korean, Russian, Sanskrit, Tamil, and Telugu. It would be interesting to consider languages from language families that apparently originate from other continents such as the Austroasiatic, Niger-Congo, Uralic, and some Native American language families.

With a larger set of language families and cognate words, it would be meaningful to construct phylogenetic trees based on the number of cognate words that can be found between each pair of language families.

Disclosure of Interests. The authors have no competing interests to declare that are relevant to the content of this article.

References

1. Bakró-Nagy, M., Laakso, J., Skribnik, E.: The Oxford Guide to the Uralic Languages. Oxford University Press, Oxford, UK (2022)
2. Beekes, R.S.P.: Etymological Dictionary of Greek. Brill, Leiden, Netherlands (2010)
3. Bouckaert, R., et al.: Mapping the origins and expansion of the Indo-European language family. Science **337**(6097), 957–960 (2012)
4. Bujas, Ž.: English-Croatian Dictionary. Globus (1999)
5. Ciobanu, A.M., Dinu, L.P.: Automatic identification and production of related words for historical linguistics. Comput. Linguist. **45**(4), 667–704 (2020)
6. Czuczor, G., Fogarasi, J.: A magyar nyelv szótára (Dictionary of the Hungarian Language). Emich Gusztáv Magyar Akadémiai Nyomda (1864)
7. De Larajasse, E.: Somali-English and English-Somali Dictionary. K. Paul, Trench, Trübner & Company (1897)
8. English, L.J.: Tagalog-English dictionary. Congregation of the Most Holy Redeemer Manila (1986)
9. Fakinlede, K.J.: Modern Practical Dictionary Yoruba-English English-Yoruba. Hippocrene Books, New York (2003)
10. Gell-Mann, M., Ruhlen, M.: The origin and evolution of word order. Proc. Natl. Acad. Sci. **108**(42), 17290–17295 (2011)
11. Gray, R.D., Atkinson, Q.D.: Language-tree divergence times support the Anatolian theory of Indo-European origin. Nature **426**(6965), 435–439 (2003)
12. Honkola, T., Vesakoski, O., Korhonen, K., Lehtinen, J., Syrjänen, K., Wahlberg, N.: Cultural and climatic changes shape the evolutionary history of the Uralic languages. J. Evol. Biol. **26**(6), 1244–1253 (2013)
13. International Phonetic Association: Handbook of the International Phonetic Association: A guide to the use of the International Phonetic Alphabet. Cambridge University Press (1999)
14. Kálmán, B.: Chrestomathia Vogulica. Tankönyvkiadó (1965)
15. Kessler, B.: Word similarity metrics and multilateral comparison. In: Proceedings of Ninth Meeting of the ACL Special Interest Group in Computational Morphology and Phonology, pp. 6–14 (2007)
16. Kolipakam, V., et al.: A Bayesian phylogenetic study of the Dravidian language family. Royal Soc. Open Sci. **5**(3), 171504 (2018)
17. Ladefoged, P.: Vowels and Consonants: An Introduction to the Sounds of Languages. Blackwell Publishers, Oxford (2000)
18. Mamo, S.: English-Maltese Dictionary. A. Aquilina (1885)

19. Miyagawa, S., DeSalle, R., Nóbrega, V.A., Nitschke, R., Okumura, M., Tattersall, I.: Linguistic capacity was present in the Homo sapiens population 135 thousand years ago. Front. Psychol. **16**(1503900) (2025)
20. Mutabazi, B., Revesz, P.Z.: A quantitative lexicostatistics study of the evolution of the Bantu language family. WSEAS Trans. Comput. **18**, 97–100 (2019)
21. Nichols, J.: Linguistic diversity in space and time. University of Chicago Press (1992)
22. Pagel, M., Atkinson, Q.D., Calude, A.S., Meade, A.: Ultraconserved words point to deep language ancestry across Eurasia. Proc. Nat. Acad. Sci. **110**(21), 8471–8476 (2013)
23. Parpola, S.: Etymological Dictionary of the Sumerian Language. Eisenbrauns, Winona Lake, Indiana, USA (2022)
24. Patyaiying, P.: Radhakrishnan's Philosophy of Religion. Gyan Publishing House, Delhi, India (2008)
25. Pellard, T., Sagart, L., Jacques, G.: L'indo-européen n'est pas un mythe. Bulletin de la Société de Linguistique de Paris **113**(1), 79–102 (2018)
26. Rédei, K. (ed.): Uralisches etymologisches Wörterbuch. Akadémiai Kiadó, Budapest, Hungary (1988)
27. Revesz, P.Z.: Establishing the West-Ugric language family with Minoan, Hattic and Hungarian by a decipherment of Linear A. WSEAS Trans. Inf. Sci. Appl. **14**, 306–335 (2017)
28. Revesz, P.Z.: Spatio-temporal data mining of major European river and mountain names reveals their Near Eastern and African origins. In: Benczúr, A., Thalheim, B., Horváth, T. (eds.) ADBIS 2018. LNCS, vol. 11019, pp. 20–32. Springer, Cham (2018). https://doi.org/10.1007/978-3-319-98398-1_2
29. Revesz, P.Z.: Using data mining algorithms to discover regular sound changes among languages. In: MATEC Web of Conferences (22nd Int. Conference on Circuits, Systems, Communications and Computers), vol. 292(03018), pp. 452–463 (2019)
30. Revesz, P.Z.: Minoan and Finno-Ugric regular sound changes discovered by data mining. In: Proceedings of the 24th International Conference on Circuits, Systems, Communications and Computers (CSCC), pp. 241–246. IEEE Press (2020)
31. Revesz, P.Z.: A vowel harmony testing algorithm to aid in ancient script decipherment. In: Proceedings of the 24th International Conference on Circuits, Systems, Communications and Computers (CSCC), pp. 35–38. IEEE Press (2020)
32. Revesz, P.Z.: Was the Uralic Homeland in the Danube Basin? Magyarok Világszövetsége, Budapest, Hungary (2021)
33. Revesz, P.Z.: Sumerian-Ugric protowords and regular sound changes, Appendix to: S. Parpola, Etymological Dictionary of the Sumerian Language, vol. 3, Eisenbrauns, Winona Lake, Indiana, USA, pp. 390–415 (2022)
34. Revesz, P.Z.: A generalization of the Chomsky-Halle phonetic representation using real numbers for robust speech recognition in noisy environments. In: Proceedings of the 27th International Database Engineered Applications Symposium (IDEAS), pp. 156–160. ACM Press (2023)
35. Revesz, P.Z.: A tale of two sphinxes: Proof that the Potaissa Sphinx is authentic and other Aegean influences on early Hungarian inscriptions. Mediter. Archaeol. Archaeom. **24**(2), 191–216 (2024)
36. Schaefer, D., Revesz, P.Z.: Comparing related languages with a fuzzy morphism matching algorithm. In: Proceedings of the 28th International Database Engineered Applications Symposium (IDEAS). LNCS, vol. 15511, pp. 79–91. Springer-Nature (2024). https://doi.org/10.1007/978-3-031-83472-1_6

37. Swadesh, M.: Lexico-statistic dating of prehistoric ethnic contacts. Proc. Am. Philos. Soc. **96**(4), 452–463 (1952)
38. Vajda, E., Fortescue, M.: Mid-Holocene Language Connections between Asia and North America. Brill, Leiden, Netherlands (2022)
39. VanOrsdale, J., Chauhan, J., Potlapally, S.V., Chanamolu, S., Kasara, S.P.R., Revesz, P.Z.: Measuring vowel harmony within Hungarian, the Indus Valley Script language, Spanish and Turkish using ERGM. In: Proceedings of the 26th International Database Engineered Applications Symposium (IDEAS), pp. 171–174. IEEE Press (2022)
40. Ventris, M., Chadwick, J.: Documents in Mycenaean Greek. Cambridge University Press (2015)
41. Yang, S., Sun, X., Jin, L., Zhang, M.: Inferring language dispersal patterns with velocity field estimation. Nat. Commun. **15**(1), 190 (2024)

Open Access This chapter is licensed under the terms of the Creative Commons Attribution-NonCommercial-NoDerivatives 4.0 International License (http://creativecommons.org/licenses/by-nc-nd/4.0/), which permits any noncommercial use, sharing, distribution and reproduction in any medium or format, as long as you give appropriate credit to the original author(s) and the source, provide a link to the Creative Commons license and indicate if you modified the licensed material. You do not have permission under this license to share adapted material derived from this chapter or parts of it.

The images or other third party material in this chapter are included in the chapter's Creative Commons license, unless indicated otherwise in a credit line to the material. If material is not included in the chapter's Creative Commons license and your intended use is not permitted by statutory regulation or exceeds the permitted use, you will need to obtain permission directly from the copyright holder.

A Cross-Linguistic Analysis of Linear A, Linear B and Swahili

Joslin Ishimwe(✉), Adrian Ratwatte, and Prince Ngiruwonsanga

School of Computing, University of Nebraska-Lincoln, 104 Schorr Center, 1100 T
Street, Lincoln, NE 68588-0150, USA
{jishimwe2,aratwatte2,pngiruwonsanga6}@unl.edu

Abstract. Prior computational work on Linear A has focused on isolated statistical models with little cross-linguistic integration. This study introduces a unified computational framework analyzing Linear A, Linear B, and Swahili through probabilistic modeling and pattern mining. We construct syllable-level Markov models, extract transition probability matrices, and apply Jaccard similarity and Apriori rule mining to uncover structural patterns. Results reveal recurring transition clusters in Linear A suggestive of morphological structure and identify significant syllabic co-occurrences in Swahili that align with patterns in ancient scripts. Cross-script analysis highlights potential phonotactic and positional correspondences, offering new computational insights into Linear A's linguistic organization. The methods presented can help in the decipherment of scripts that lack bilingual texts.

Keywords: Ancient scripts · Morphological analysis · Symbol transitions · Cross-script association

1 Introduction

Efforts to decipher ancient scripts have long fascinated linguists, historians, and computational researchers alike. Among the most enigmatic of these scripts is Linear A, a writing system used in Minoan civilization that remains undeciphered to this day [7]. While its successor, Linear B, was successfully decoded in the1950 s and shown to represent an early form of Greek, Linear A's linguistic classification is still unknown, despite structural similarities in its glyphs [8]. Existing studies, such as Michele et al. [2], have employed statistical and probabilistic approaches to explore script regularities. However, these studies have predominantly focused on either a single script or monolingual comparisons, leaving a significant gap in cross-linguistic computational analyses involving modern spoken languages. Furthermore, although some work has applied Markov models to undeciphered scripts, few have integrated advanced similarity metrics or multilinguistic corpora into a unified framework [15].

This paper addresses this research gap by presenting a cross-linguistic computational analysis of Linear A, Linear B, and Swahili using a unified probabilistic framework. Specifically, we employ Markov models to model intra-script syllabic transitions, and we extend the analysis by applying Jaccard similarity and Apriori-based frequent pattern mining across scripts. Swahili, as a well-documented agglutinative language with a rich syllabic structure [15], serves as a modern linguistic comparator that can potentially illuminate recurring phonological or grammatical structures in ancient scripts. Our work expands upon prior efforts by not only building transition matrices for each language but also comparing them quantitatively through cross-script metrics. This integrative approach allows us to identify potential phonotactic or morphological alignments that could offer new clues into the function and structure of Linear A.

The main goals of this research are: (1) to construct syllable-level probabilistic models for Linear A, Linear B, and Swahili; (2) to develop computational tools for cross-script structural comparison using transition probability matrices and Jaccard similarity indices; and (3) to extract frequently co-occurring syllabic sequences via Apriori association rule mining [1].

The paper is organized as follows: Sect. 2 details the Linear A modeling methodology, covering data preparation, transition matrices, and visual analysis. Section 3.1 describes the Swahili corpus, syllabic segmentation, and Markov modeling for morphological insights. Sections 3.2 and 3.3 present the cross-script comparison using associative rule mining and Jaccard similarity. Section 4 presents some conclusions.

2 Linear A, Linear B, Swahili

Linear A is a syllabic script from Minoan civilization that existed on island of Crete between 1450 and 1200 BCE. Our work in this paper aims to uncover the structural properties of Linear A using statistical modeling and comparison with Linear B and Swahili. Swahili, a Bantu language with agglutinative morphology and rich syllabic structure, presents unique challenges and opportunities for computational linguistics. Despite increasing digital content, detailed probabilistic models of Swahili syllable transitions remain under-explored. Section 3 introduces a syllable-level Markov modeling framework tailored to Swahili, enabling statistical analysis and visualization of transition patterns. Linear A has several decipherment proposals, including a recent proposal as a language belonging to the Ugric branch of the Uralic language family [11]. This proposal is supported by archaeogenetics that connects the Minoans with Uralic peoples [13].

Linear B is a syllabic script used to write Mycenaean Greek, which represents the earliest known form of the Greek [5]. It was used primarily for administrative and economic records during the Late Bronze Age, specifically around 1450–1200 BCE. Linear B was used on Crete and the Peloponnese. It is notable as the earliest deciphered form of Greek, predating the Greek alphabet. Linear B was ingeniously deciphered by Michael Ventris in 1952 [5]. Linear A and Linear B belong to the same cluster of Bronze Age scripts [3], which is also called the Cretan Script Family [10]. The script utilized around 87 syllabic signs, such as *po*, *te*, and *ra*, and over 100 ideograms. Records written in Linear B typically consist of text, an ideogram, and a number. These records, found at sites like Pylos and

Knossos, provide significant insight into the goods and products available at the time [5].

Swahili belongs to the Bantu language family [6] and is spoken in eastern Africa. Swahili is interesting to investigate because there was an ancient population migration from eastern Africa to Europe that is reflected in many similar toponyms in Africa, the Near East and Europe [12].

2.1 Data Preparation

We found around 7400 Linear A signs distributed across 1527 inscriptions mostly from clay tablets, roundels, and nodules [4]. We extracted raw inscriptions from *GORILA* corpus (Godard & Olivier, 1976–1985) and John G. Younger's Linear A database. To prepare our data, we selected syllabified words only, removed fragmented or uncertain entries from ~1100 Linear A words found. The resulting inscriptions were tokenized into syllables (i.e. "A-KA-RU" → ["A", "KA", "RU"]). After processing the data, the final input list had around 700 clean syllabified Linear A sequence words ready for modeling. For a full list of Linear A syllabified words refer to Linear A decipherment programme by Dr. Francesco Perono Cacciafoco (NTU, Singapore) [14].

For the Linear B portion of the analysis, the data was sourced from the website https://linear-b.kinezika.com/en/. This source provided a dataset that initially contained approximately ~2500 words. The data processing involved several cleaning and standardization steps before the data was ready for modeling. Entries lacking valid transcriptions were removed, as were any syllabic sequences containing numbers. All transcription entries were also standardized by converting them to uppercase. Finally, like the Linear A data, the transcriptions were tokenized by splitting on hyphens to obtain individual syllabic tokens; for example, a transcription such as 'ri-ta-ro' was converted to ["RI", "TA", "RO"]. This prepared list of syllabic sequences was then ready for computational analysis, such as modeling syllable transitions using a Markov model, as outlined in the project goals.

We curated a Swahili health-related corpus [9] and conducted systematic preprocessing. The text was lowercased, punctuation was removed, and whitespace-separated tokens were segmented into syllable-like chunks using a custom rule that splits longer words into trigrams, merging short trailing fragments to preserve phonological plausibility. This generated a sequence of syllabic units suitable for probabilistic modeling.

2.2 Algorithm Implementation

Markov Model: To analyze Linear A and Linear B, we started with the Markov model. Our Markov model is a transition system from states to states, where each state is a script sign. We used the Markov model to measure the transition probability represented by the likelihood that one syllable follows another. We constructed transition matrices to capture relationships between syllabic signs.

The transition matrices were used to generate a heatmap visualization using seaborn in Python to demonstrate frequent sign transitions.

To develop the Markov model for Swahili, we preprocessed the raw corpus AFYA_Cleaned.txt by lowercasing, removing punctuation, and tokenizing into words. Words longer than three characters were split into 3-letter syllabic chunks, merging short trailing segments to maintain phonological coherence. The resulting syllable sequence was used to build a first-order Markov model, generating a transition count matrix normalized into transition probabilities. Top syllables were segmented into batches to visualize patterns and compute Jaccard similarity, revealing structural and phonological relationships in the syllabic flow.

Jaccard Similarity Analysis: After the Markov model, we performed cross-script analysis using Jaccard Similarity. Jaccard similarity quantified the structural overlap between Linear A, Linear B, and Swahili using shared n-grams. The process involved extracting bigrams from the Markov model of each script, then measuring the overlap between two-set of n-grams using the formula below:

$$J(A, B) = \frac{|A \cap B|}{|A \cup B|} \quad (1)$$

Association Rule Mining: This study employs a systematic approach to extract and analyze consonantal patterns from syllabified Swahili, Linear A, and Linear B data. Initial (ini), medial (med), and final (fin) consonants are extracted per language and merged into a unified dataset. One-hot encoding is applied to convert syllables into transactions for Apriori-based association rule mining. Frequent item sets are identified using a minimum support threshold, and rules are filtered by confidence. A selection of representative rules for each script is presented in the next section.

3 Results and Discussion

3.1 Markov Model

We use the Markov model to compute the transitional matrices, cluster analysis, and network graphs for each script. The results showed novel organizational features and syllabic patterns across Linear A, Linear B, and Swahili scripts. In this section, we will represent our finding in two main categories: (1) transition probability analysis, (2) cross-script comparisons using Jaccard similarity and association rule mining.

To understand the sequential dependencies in Linear A and Linear B, we constructed transition probability matrices from cleaned syllabic sequences discussed above. This generated row count heatmaps that visually represent the syllabic transitions within each script.

Figure 1 presents the transition count matrix for the top 40 Swahili syllables, capturing raw frequencies of sequential syllables. High counts such as a → a and

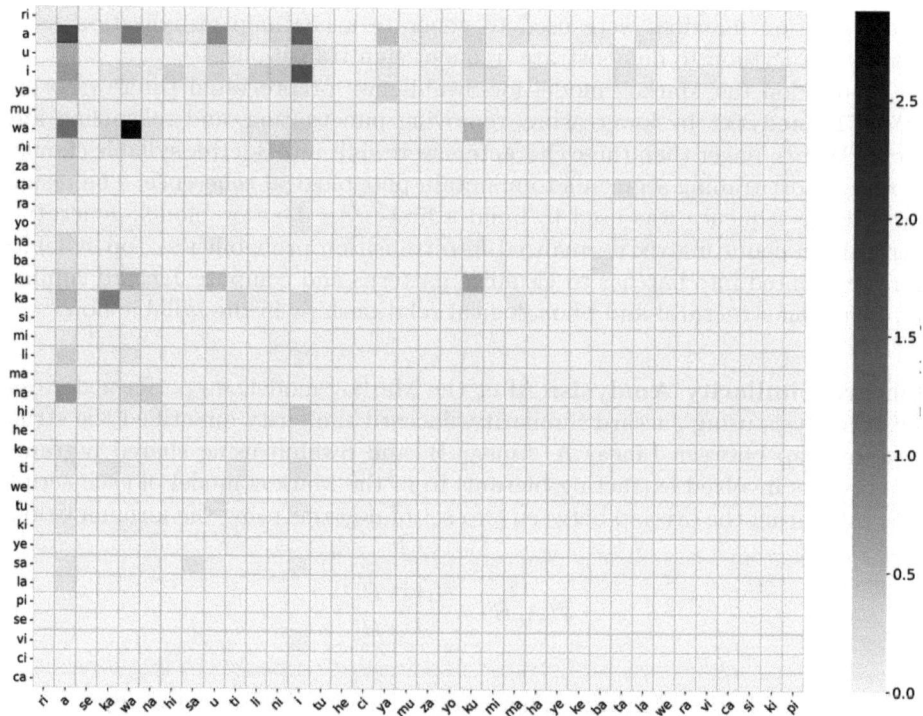

Fig. 1. Transition count matrix of Swahili syllables (batch 1).

i → i point to common verb constructions, while lower-frequency transitions help trace less dominant but structurally relevant patterns in the language.

Figure 2 illustrates the corresponding raw count matrix for Linear B. Transitions such as RI-JO (54 times), RE-U (49 times), KA-RA (46 times), TO-RO (39 times), and SI-JO (35 times) stood out on the syllable transition matrix. The frequency of these transitions indicates syllable pairs that are likely essential to the morphology or structure of Linear B inscriptions, most likely corresponding to the standard administrative terminology or commonly referenced institutions.

3.2 Association Rule Mining for Swahili, Linear A and Linear B

Table 1 reveals that final-position symbols in Linear A, such as RO (fin), exhibit strong associations with a consistent set of symbols across different positions, indicating their potential role as structural anchors or grammatical markers. In contrast, medial symbols like SU (med) and RO (med) show weaker, more variable associations, suggesting they serve more flexible or syntactic functions within inscriptions. This pattern points to a positional hierarchy in Linear A, where final symbols likely help structure or delimit meaningful units, while medial elements provide internal variation.

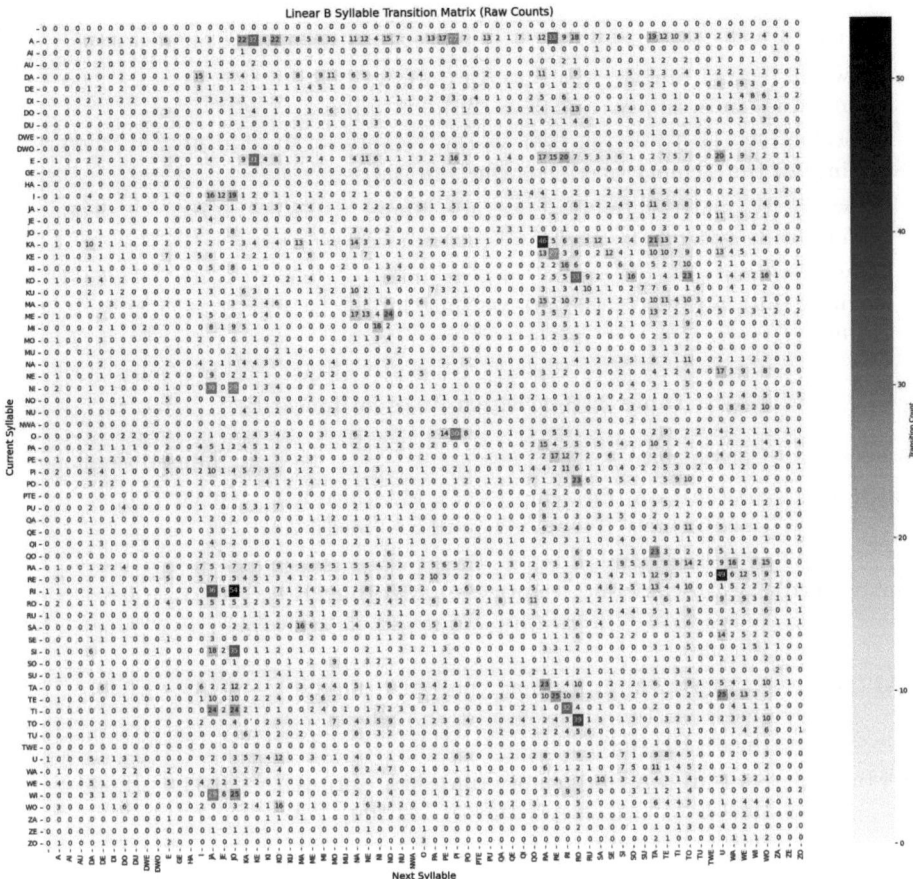

Fig. 2. Transition count matrix of Linear B syllables.

Table 1. Association rules for Linear A.

Antecedents	Consequents	Support	Confidence
RO (fin)	KI (ini), SU (med), RO (med), DI (med), ZU (ini), DI (ini), RA (med), ZE (fin), QE (ini), SU (fin), U (fin)	0.50	0.8
SU (med)	KI (ini), RO (fin), RO (med), DI (med), ZU (ini), DI (ini), RA (med), ZE (fin), QE (ini), SU (fin), U (fin)	0.2	0.5
RO (med)	KI (ini), RO (fin), SU (med), DI (med), ZU (ini), DI (ini), RA (med), ZE (fin), QE (ini), SU (fin), U (fin)	0.1	0.2

Table 2. Association rules for Swahili.

Rule	Support	Confidence	Lift
Swahili (empty (fin)) ⇒ (na (ini), va (ini), wa (med), wa (ini), empty (med), wa (fin), ku (ini))	0.50	1.00	2.00
Swahili (wa (med)) ⇒ (na (ini), va (ini), empty (fin), wa (ini), empty (med), wa (fin), ku (ini))	0.50	0.40	3.00
Swahili (wa (ini)) ⇒ (na (ini), va (ini), empty (fin), wa (med), empty (med), wa (fin), ku (ini))	0.50	0.10	5.00
Swahili (empty (med)) ⇒ (na (ini), va (ini), empty (fin), wa (med), wa (ini), wa (fin), ku (ini))	0.50	0.70	6.00

Table 3. Association Rules for Linear B.

Antecedent	Consequent	Support	Confidence
AI (ini)	SA (fin), ZA (fin), empty (fin), MA (fin), KA (med), empty (med), "AI (ini), SA (med), ZA (med), PE (ini)	0.33	0.98
empty (med)	SA (fin), ZA (fin), empty (fin), MA (fin), KA (med), AI (ini), "AI (ini), SA (med), ZA (med), PE (ini)	0.33	0.8
"AI (ini)	SA (fin), ZA (fin), empty (fin), MA (fin), KA (med), AI (ini), empty (med), SA (med), ZA (med), PE (ini)	0.33	0.85
SA (med)	SA (fin), ZA (fin), empty (fin), MA (fin), KA (med), AI (ini), empty (med), "AI (ini), ZA (med), PE (ini)	0.33	1.00

The Swahili rules suggest that certain morphemes, particularly "wa" and "empty" tokens in initial, medial, and final positions, are highly interconnected and frequently co-occur with core grammatical markers like "na," "ya," and "ku." as shown in Table 2. The consistent presence of "empty (fin)" across rules with perfect confidence and high support indicates it may function as a structural or grammatical boundary, possibly marking phrase or sentence endings. Additionally, the repeated centrality of "wa" across all positions implies it plays a versatile grammatical role, such as marking subjects or objects. These patterns highlight a strong positional dependency and suggest that Swahili morphology relies on predictable co-occurrences, reinforcing its agglutinative structure. Table 3 summarizes the association rules derived for the Linear B dataset using Apriori analysis. Each rule exhibits perfect confidence above 0.8 and a high lift value (3.00), indicating strong and reliable associations. The consistent presence of terminal states such as SA (fin), ZA (fin), and MA (fin) across all consequents suggests a dominant and predictable outcome pattern. Antecedents such as AI

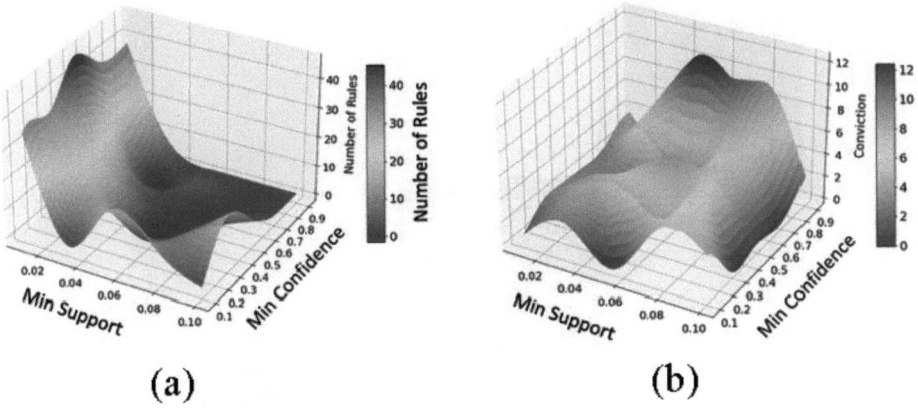

Fig. 3. Apriori association rule mining: (a) Rule count across varying minimum support and confidence thresholds. (b) Conviction values for the same threshold combinations.

(ini), SA (med), and ZA (med) appear repeatedly, highlighting their influential role in driving final decisions.

Figure 3(a) demonstrates the number of discovered rules increases rapidly at lower minimum support thresholds (0.02–0.04), especially when paired with moderate confidence levels, indicating that many meaningful but less frequent patterns emerge in this range. Conversely, higher confidence thresholds (above 0.7) yield fewer but stronger rules with greater predictive power, as evidenced by higher in lift and conviction scores (Fig 3(b)). This trade-off suggests that effective rule mining requires careful calibration of support and confidence to balance rule richness with reliability. These findings complement the earlier observation that structural elements like "empty (fin)" and the versatile use of "wa" dominate the rule space, confirming both the predictability and the functional complexity of Swahili's morphological system.

3.3 Jaccard Similarity

Our Jaccard similarity graph (Fig. 4) reveals an interesting pattern in structural relationships between Linear A, Linear B, and Swahili. The bigrams were first normalized-capital letters only- to ensure that all transitions are structurally similar. We found the strongest similarity to be between Linear B and Swahili ($J = 0.0865$), followed by Linear A and Linear B (0.0470), and finally the weakest similarity was between Linear A and Swahili (0.0198). These measurements do not suggest that Swahili, Mycenaean Greek, and the underlying language of Linear A are related. We also captured bigrams similarity between all three scripts to uncover which bigrams are similar across all three. with 88 shares bigrams. Linear B and Swahili shared 209 bigrams, Linear A and Linear B shared 88 bigrams, and Linear A and Swahili shared 19 bigrams. The similarity between Linear B and Swahili was very unexpected and it appears to be through various

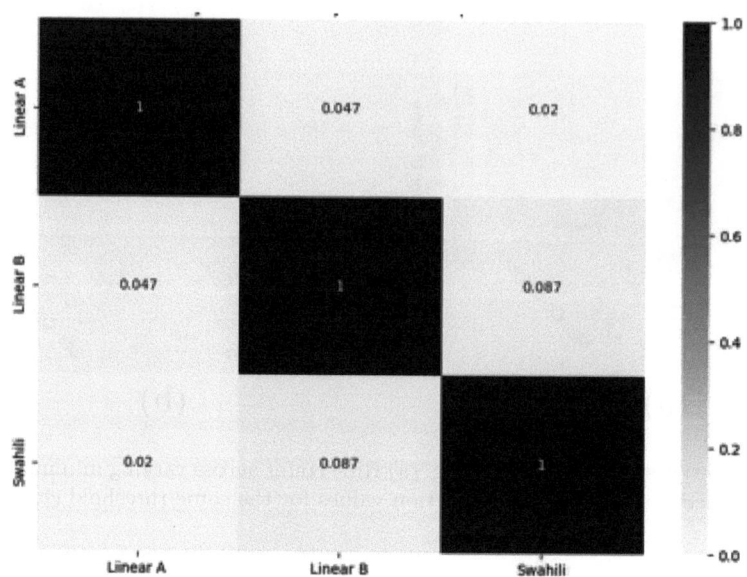

Fig. 4. Jaccard similarity between Linear A, Linear B, and Swahili.

shared consonants-vowel syllabic structures such as KI-TI, A-MU, TI-WA, A-WA, and KE-NA and similar positional distributions. This similarity is likely an indication of universal tendencies in syllabic script design rather than a direct connection historically between Linear b and Swahili. However, while these similarities may appear smaller, they are statistically significant and would prove to be an interesting area for further research. Linear B and Swahili similarity index may suggest that some conventions are common in language design.

4 Conclusion

This study introduces a unified probabilistic framework for analyzing Linear A, Linear B, and Swahili, enabling both intra-script modeling and cross-linguistic comparison. Using syllable-level Markov models, Jaccard similarity, and Apriori-based rule mining, we identify structural motifs and transition patterns suggesting syntactic or phonological parallels. The inclusion of Swahili, a modern agglutinative language, offers a comparative lens for contextualizing regularities in ancient scripts.

Our findings show that script relationships can be continuous rather than binary. Despite no clear genealogical ties, Jaccard similarity proved effective in quantifying structural parallels. This framework lays groundwork for future cross-script computational linguistic studies. Future work will extend the corpus with annotated data and explore phonological embeddings and neural models to enhance transition interpretability. We also aim to investigate cross-script association rules to reveal deeper structural or semantic links across distinct writing systems. The source code and examples for Markov modeling and association rule mining across Linear A, Linear B, and Swahili are available on GitHub: https://github.com/ishjosl/LinearAB-Swahili-Decipherment

Disclosure of Interests. The authors have no competing interests to declare that are relevant to the content of this article.

References

1. Borgelt, C., Kruse, R.: Induction of association rules: apriori implementation. In: Compstat: Proceedings in Computational Statistics. pp. 395–400. Springer (2002). https://doi.org/10.1007/978-3-642-57489-4_59
2. Corazza, M., Ferrara, S., Montecchi, B., Tamburini, F., Valério, M.: The mathematical values of fraction signs in the linear a script: a computational, statistical and typological approach. J. Archaeol. Sci. **125**, 105214 (2021)
3. Daggumati, S., Revesz, P.Z.: Convolutional neural networks analysis reveals three possible sources of Bronze Age writings between Greece and India. Information **14**(227) (2023)
4. Del Freo, M., Zurbach, J.: La préparation d'un supplément au recueil des inscriptions en linéaire a. observations à partir d'un travail en cours. Bulletin de Correspondance hellénique **135**(1), 73–97 (2011)
5. Duan, F., Sun, M., Chen, X., Jiang, W.: Ovarian clear cell carcinoma: research progress in oncogenesis and novel therapeutic strategies. Academia Oncol. **1**(1) (2024)
6. Mutabazi, B., Revesz, P.Z.: A quantitative lexicostatistics study of the evolution of the Bantu language family. WSEAS Trans. Comput. **18**, 97–100 (2019)
7. Packard, D.W.: Minoan Linear A. Univ of California Press (1974)
8. Palaima, T.G.: Linear b. The Oxford Handbook of the Bronze Age Aegean, pp. 356–72 (2010)
9. Rao, R.P., Yadav, N., Vahia, M.N., Joglekar, H., Adhikari, R., Mahadevan, I.: A markov model of the indus script. Proc. Natl. Acad. Sci. **106**(33), 13685–13690 (2009)
10. Revesz, P.Z.: Bioinformatics evolutionary tree algorithms reveal the history of the cretan script family. Intern. J. Appli. Math. Inform. **10**(1), 67–76 (2016)
11. Revesz, P.Z.: Establishing the West-Ugric language family with Minoan, Hattic and Hungarian by a decipherment of Linear A. WSEAS Trans. Inf. Sci. Appl. **14**, 306–335 (2017)
12. Revesz, P.Z.: Spatio-temporal data mining of major european river and mountain names reveals their near eastern and African origins. In: Benczúr, A., Thalheim, B., Horváth, T. (eds.) ADBIS 2018. LNCS, vol. 11019, pp. 20–32. Springer, Cham (2018). https://doi.org/10.1007/978-3-319-98398-1_2
13. Revesz, P.Z.: Archaeogenetic data mining supports a Uralic-Minoan homeland in the Danube Basin. Information **15**(10, article number 646) (2024). https://doi.org/10.3390/info15100646
14. Thanopoulos, R., Drossinou, I., Koutroumpelas, I., Chatzigeorgiou, T., Stavrakaki, M., Bebeli, P.J.: Hilly, semi-mountainous and mountainous areas harbor landraces diversity: the case of messinia (peloponnese-greece). Diversity **16**(3), 151 (2024)
15. Walsh, M.: The swahili language and its early history. In: The Swahili World, pp. 121–130. Routledge (2017)

Automated Identification of Allographs Among the Indus Valley Script Signs

Harsh Tamkiya, Gunjit Agrawal, Chiradeep Debnath, and Peter Z. Revesz(✉)

School of Computing, University of Nebraska-Lincoln, Lincoln, NE 68508, USA
{htamkiya2,gagrawal2,cdebnath2,prevesz1}@unl.edu

Abstract. One of the major reasons for the lack of decipherment of the Indus Valley Script is that even its set of signs have not been precisely identified. This paper proposes a fully automated method leveraging computer vision and statistical modeling to analyze the Indus Valley Script. The automated method includes sign segmentation using adaptive thresholding and morphology, followed by visual feature extraction with VGG16 deep learning and dimensionality reduction via principal component analysis. Clustering with K-means grouped the previously proposed 417 Indus Valley Script signs into 50 clusters that would be a more reasonable number if the Indus Valley Script is a syllabic or alphabetic script. The automated method also builds a first-order Markov chain on the 50 sign clusters. The Markov model reveals some frequent self-loops and other interesting patterns that hint at the grammar of the underlying language of the Indus Valley Script. The grammar could lead to the identification of related languages, aiding the decipherment of the Indus Valley Script. The proposed automated method could be adapted to the study of other undeciphered scripts.

Keywords: Allograph · Clustering · Markov model · Segmentation · Indus Valley Script

1 Introduction

The Indus Valley Civilization, which is also called the Harappan Civilization, existed from about 2600 to 1900 BCE in what is now Pakistan, northwest India, and south-eastern Afghanistan. One of the most interesting features of the Indus Valley Civilization is its still undeciphered writing system, which is known as the Indus Valley Script [8].

While over four thousands inscriptions have been found on seals, tablets and pottery fragments and listed in the *Corpus of Indus Seals and Inscriptions* by Joshi and Parpola [6], none of those inscriptions is a bilingual inscription like the famous "Rosetta Stone" that helped to decipher the Egyptian hieroglyphs. Moreover, the inscriptions are usually short sequences of signs with the longest inscription containing only 34 signs [6].

A manual analysis of the signs led Mahadevan [8] to identify 417 different signs and Wells [18] to identify 694 signs in the Indus Valley Script. Not only are these different identifications contradictory and confusing, but also they are too high numbers if the script is a syllabary like the Sumerian Pictograms to which it is closest related among the Bronze Age scripts as recently shown by Daggumati and Revesz [3] using a convolutional neural networks analysis. We would expect to have about 80 signs in a syllabary. Moreover, if it is an alphabet, then we would expect it to have about 50 signs because even phonetically complex languages tend to have about as many different phonemes. Clearly, there is a huge gap between the 417 different signs that have been identified and frequently cited in the literature of the Indus Valley Script, and the 80 or 50 signs that we would expect in the case of syllabic or alphabetic scripts, respectively.

The natural research question that arises from the above consideration is how to reduce the 417 different signs to about 50 signs. That is a fundamental question because any serious decipherment can only begin once the different signs are solidly identified. Even then we would have a huge search space trying to identify which sign denoted which syllable or phoneme, but certain AI techniques could cut down the huge search space and may yield a solution in this step.

Daggumati and Revesz [2] showed that many of the similar-looking Indus Valley Script signs are allographs of each other, that is, they have the same phonetic values. Many of these allographs look very similar to each other and their minor differences could be explained by an apparent intention of the scribe to save space on the small seals. For example, a bow sign that is normally curved like a right parenthesis) could be written after an X-like sign as a left parenthesis (to save space [2] because the left parenthesis could fit into the right side of the X-like sign when writing X(.

Daggumati and Revesz [2] could manually identify about 50 pairs of signs to be allographs. Thereby, they reduced the 417 different signs to about 367 signs. Unfortunately, manual clustering could not take further the reduction of the 417 signs that Mahadevan [8] identified. Hence, we now apply some advanced computer techniques to the Indus Valley Script signs as a way to make further progress. We describe and implement a complete software system that does the following three things:

1. Automatically separates and cleans the signs from the pictures of the inscriptions.
2. Groups similar-looking signs using deep learning and clustering,
3. Builds a Markov model on the clusters, which helps identify some patterns that maybe correlated with certain grammatical features.

The rest of this paper is organized as follows. Section 2 describes our data processing method to generate the sign sequences from the pictures of the inscriptions. There is a need to do this using a computer vision software because there are many errors in the manual entries as noted by Daggumati and Revesz [2]. Section 3 describes our feature extraction and clustering methods. Section 4 describes building a Markov model for the clusters. Section 5 discusses the results and provides some analysis of the interesting patterns that emerge from the

Markov model. Finally, Sect. 6 gives some conclusions and mentions possible future work.

2 Data and Preprocessing

We used a mix of real Indus Valley Script seal images and cleaned-up sign images from existing datasets.

2.1 Datasets Used

We collected 130 high-quality photos of actual Indus seals from different archaeological sites like Harappa, Lothal, and Mohenjo-daro. These images were taken from public sources like Harappa.com [5] and a collection by Fuls [4].

In addition, we used a labeled dataset called IM-417-150 from Kaggle [7], which contains 417 individual signs. These signs have already been separated from the seal images and can be used to train and test machine learning models.

Using both types of data helped us build a system that can recognize individual signs on seals and group them based on how they look.

2.2 Image Enhancement

The original seal photos were not clean or uniform. Some were blurry, some had low contrast, and most had extra background details we did not need. To fix these problems, we applied several image processing steps:

- Grayscale conversion: Removed color and focused only on brightness levels.
- Contrast stretching: Improved the difference between light and dark areas.
- Noise reduction: Used a filter to smooth out grainy areas.
- Adaptive thresholding: Separated signs from the background based on local brightness.
- Morphological operations: Used opening and closing techniques to clean up shapes and remove small unwanted parts.

These steps made the signs easier to detect and process.

2.3 Sign Isolation

After cleaning up the seal images, we needed to find and extract each individual sign. We used a technique called connected components, which groups together pixels that belong to the same shape.

We also added some filters:

- Size filter: Ignored shapes that were too small or too large.
- Aspect ratio filter: Ignored shapes that were too long or too wide to be a proper sign.

To make sure the system was working, we checked some of the results by hand. Around 85% of the signs were correctly detected. Some difficult cases like overlapping signs or broken images will be handled using more advanced tools like deep learning segmentation in future work.

3 Feature Extraction and Clustering

After extracting individual signs from the seal images, the next step was to find a way to compare them. Since these are ancient signs with many different shapes and styles, we used machine learning to turn each sign into a set of numbers that represent how it looks. Then we grouped similar signs together using clustering.

3.1 Deep Feature Vectorization

Each sign was resized to a standard size of 224×224 pixels. We then passed it through a deep learning model called VGG16, which is a popular convolutional neural network trained on image recognition tasks. We used the model without the final classification layer because we did not want to label the signs. We just wanted to get their visual features.

The output of VGG16 for each image was a $512 \times 7 \times 7$ activation tensor, which we flattened into a vector with 25,088 values. That gave us a detailed description of each sign's appearance.

3.2 PCA Reduction

Since 25,088 is too many for clustering, we used a method called Principal Component Analysis (PCA) to shrink the feature vectors while keeping most of the important information.

After applying PCA, we reduced each sign's vector to just 50 values, which still captured around 92% of the total variance. That made it easier and faster to group the signs, while still keeping the key differences between them.

Figure 1 shows how much information is kept by each principal component. We noticed a clear elbow in the plot, meaning that adding more than 50 components did not add much extra value.

3.3 K-Means Clustering

With the reduced 50-dimensional vectors, we used K-means clustering to group similar signs. We tested different values of K and found that $K = 50$ gave the best results based on two measures:

- Inertia: How tightly grouped each cluster is.
- Silhouette score: How well-separated each cluster is from other clusters.

We used K-means++ to improve the quality of the initial cluster centers. Each of the 417 known signs was assigned to one of these 50 clusters. Some clusters had clear patterns like signs with loops or dots, while others had more complex shapes. Figure 2 shows some sample clusters. Cluster 4 groups together those signs which have several vertical bars. Cluster 18 groups together signs that have an X or a diamond-like shape. Cluster 45 groups together signs that are variations on two partially overlapping ovals. The clustering cannot tell us what

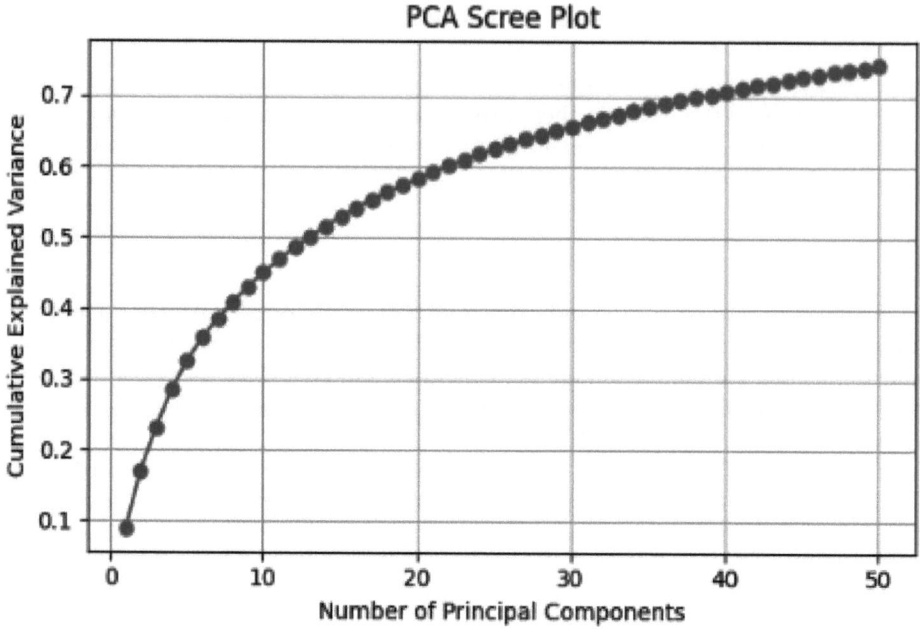

Fig. 1. PCA explained by the variances of the components.

these signs mean. The clustering only suggests that they are allographs, that is, they denote the same phoneme. However, it is also possible that in Cluster 4, the signs denote various numbers where the number of vertical bars indicates the denoted number. For example, IIII would denote the number 4. This notation is different from the Roman numeral IV, which suggests that the Indus Valley Civilization did not have a close connection with Roman culture and the Latin language.

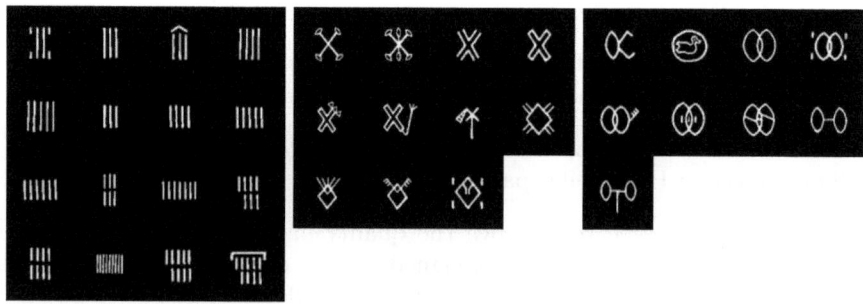

Fig. 2. Sample clusters: (left) Cluster 4, (center) Cluster 18, (right) Cluster 45.

4 The Markov Modeling Results

Once we grouped similar signs into clusters, the next step was to look at how these signs appear in sequence. Even though we do not know what the signs mean, we can still study the order in which they are used. This can help us understand whether there is any grammar or structure in the Indus script.

We used a first-order Markov model to analyze these sequences. A Markov model usually describes how likely it is for one sign to follow another. Rao et al. [9] already applied Markov models to the 417 Indus Valley Script signs identified by Mahadevan [8], but that did not give results that led any closer to a decipherment. Instead, we use the Markov model to describe how likely it is for the signs in a cluster to follow the signs in another cluster.

4.1 Sequence Encoding

Each seal contains a short sequence of signs. Following to standard practice, we read these sequences from right to left, as most experts believe this was the writing direction used in the Indus Valley Script.

Instead of using the original signs, we used the cluster IDs assigned during the K-means clustering step. For example, the inscription of the seal in Fig. 3 would be considered the sequence from right-to-left: 45, 18, 4 because the rightmost sign belongs to Cluster 45, the middle sign belongs to Cluster 18, and the leftmost sign belongs to Cluster 4.

Fig. 3. Sample seal inscription.

4.2 Transition Matrix

We went through all the seal sequences and counted how often each cluster was followed by another. These counts were used to build a transition matrix where each row represents the current cluster of signs, and each column represents the next cluster of signs in the sequence.

We then normalized each row to get probabilities. For example, if Cluster 8 is followed by Cluster 22 in 12% of the cases, that entry in the matrix becomes 0.12. The formula used for the row normalization was the following:

$$P_{ij} = \frac{\#(i \to j)}{\sum_k \#(i \to k)}.$$

4.3 Visualization

Figure 4 shows a heatmap, which is a visualization of matrix P. The darker colors represent higher and the lighter colors represent lower probabilities in the heatmap. According to the heatmap in Fig. 4, the following transitions have the highest probabilities:

- The transitions from Clusters 4, 18, 24, 29, and 45 to Cluster 4.
- The transitions from Clusters 4, 18, 21, 24, 33, and 45 to Cluster 18.
- The transitions from Clusters 4, 9, 18,21, and 45 to Cluster 45.

5 Discussion of the Results

The Markov model shows some interesting patterns in the sequences of signs on the Indus Valley Script inscriptions. While we do not know what the signs mean, their arrangement suggests some kind of grammatical rules. Here are the most salient patterns.

First, some clusters have strong self-loops, that is they are very likely to repeat themselves. These include the following clusters:

- Clusters 4, 18 and 45 have strong self-loops.

These cluster self-loops suggest either geminate consonants, if the signs denote consonants, or repeated syllables like in the word *banana*, where the second and third syllables are the same. A high frequency of repetitive syllables maybe a feature of the underlying language of the Indus Valley Script.

Second, the same three clusters which have a strong self-loop also tend to follow many other clusters.

- Cluster 4 frequently follows four other clusters.
- Cluster 18 frequently follows five other clusters.
- Cluster 45 frequently follows four other clusters.

It is possible to interpret Clusters 4, 18 and 45 as some kind of grammatical markers. For example, they may denote masculine, feminine and neutral grammatical genders. This may be useful information in searching for a language that may be a relative of the language of the Indus Valley Civilization. For example, both Sanskrit and Tamil have three grammatical genders, but Vietnamese has no grammatical gender. Hence, Sanskrit and Tamil are both possible relatives, but Vietnamese is not a possible relative of the language of the Indus Valley Civilization if these three clusters mark grammatical gender.

Third, although there are 50 clusters, over 85% of all transition entries lie below 0.02. This extreme sparsity is also a characteristic of some languages. Hence, the consideration of this feature also helps to narrow down the set of possible relatives of the language of the Indus Valley Civilization.

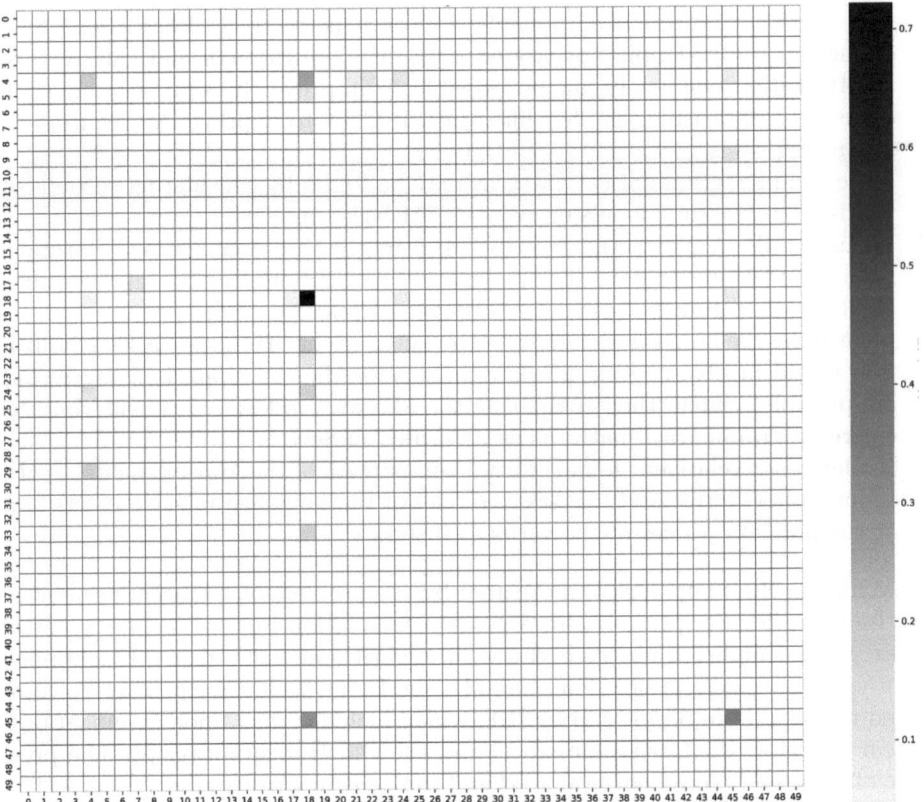

Fig. 4. Heat-map of cluster-to-cluster transition probabilities P_{ij}. The transitions are from the row cluster to the column cluster identified by the identification numbers from 0 to 49. The scale on the right side shows the transition probability.

6 Conclusions and Future Work

We created an automated system to study the Indus Valley script using modern computer vision and sequence modeling techniques. Starting with raw images of seals, we developed a pipeline that:

1. Extracts and cleans individual signs.
2. Groups similar-looking signs using deep learning and clustering.
3. Analyzes the order of signs using a Markov chain.

Our results show that the sign sequences of the Indus Valley Script inscriptions are not random. We found some strong self-loops and other interesting patterns that suggest some kind of grammar and may help narrow down the possible living relatives of the language of the Indus Valley Civilization. Some ways to improve and extend our work in the future including the following.

1. Our current connected-components approach correctly isolates about 85% of the signs. Overlapping or eroded signs still slip through and can perturb the transition counts. It maybe possible to improve the sign segmentation by using deep learning models like Mask R-CNN to improve the accuracy in separating signs.
2. More complex grammatical patterns may be found by considering second- or third-order Markov models or even LSTM-based models.
3. We used Indus Valley Script inscriptions from the three main sites of Harappa, Lothal, and Mohenjo-daro. The inclusion of inscriptions from smaller sites like Dholavira and Rakhigarhi could improve the Markov model.
4. Identifying the language relatives of the language of the Indus Valley Civilization would involve making Markov models for the candidate languages that are possible related, then comparing the Markov models and considering the emerging grammatical patterns as described in Sect. 5.
5. If a language relative is found, then a comparison of the Markov model of the Indus Valley Script and a Markov model of the language relative could hint at the phonetic values of the various clusters.

While much of our work is motivated by the problem of the decipherment of the Indus Valley Script, it is also a general foundation for the decipherment of other yet undeciphered scripts because our algorithmic method could be applied to other sets of inscriptions. For example, the Indus Valley Script may be compared to the Cretan Script Family [10], which includes the Minoan civilization's Linear A script, the Carian alphabet, and the Old Hungarian script. The presence of front-back vowel harmony [12] and the percentage of symmetric signs [16] were investigated within the Cretan Script Family. A preliminary investigation of front-back vowel harmony in the underlying language of the Indus Valley Script [17] showed some weak similarities under the assumption that the Indus Valley Script signs denote consonant-vowel pairs.

Writing could have spread between India and the Aegean region using Bronze Age trade routes that went via the Black Sea [13], where the Carians founded many colonies later. The Hungarians likely learned the use of the Carian alphabet from these colonies by the third century [15]. The Carian alphabet later developed into the Old Hungarian script [11]. That is possible because the common ancestors of the Hungarians and the Minoans lived near the lower Danube Basin and the western Black Sea regions and separated only at the beginning of the Bronze Age, when the ancestors of the Minoans migrated to Crete while the Hungarians may have remained near the Black Sea [14]. The Carian alphabet was recently deciphered [1] and the Old Hungarian script is also understood. Hence, these scripts could help in the decipherment of the Indus Valley Script.

Disclosure of Interests. The authors have no competing interests to declare that are relevant to the content of this article.

References

1. Adiego, I.J.: The Carian Language, Handbook of Oriental Studies, vol. 73. Brill, Leiden (2007)
2. Daggumati, S., Revesz, P.Z.: A method of identifying allographs in undeciphered scripts and its application to the Indus Valley Script. Humanities Soc. Sci. Commun. **8**(50) (2021)
3. Daggumati, S., Revesz, P.Z.: Convolutional neural networks analysis reveals three possible sources of Bronze Age writings between Greece and India. Information **14**(article number 227) (2023). https://doi.org/10.3390/info14040227
4. Fuls, A.: Corpus of Indus Inscriptions (2023), unpublished Manuscript
5. Harappa.com: Resources and imagery for the Indus Valley civilization (2024). https://www.harappa.com, Accessed April 2024
6. Joshi, J.P., Parpola, A.: Corpus of Indus Seals and Inscriptions. Archaeological Survey of India (1987)
7. Kaggle: Im-417-150 Indus Script Sign Dataset (2024). www.kaggle.com/datasets/storesource/im-417-150, Accessed April 2024
8. Mahadevan, I.: The Indus Script: Texts. Concordance and Tables, Archaeological Survey of India (1977)
9. Rao, R.P., Yadav, N., Vahia, M.N., Joglekar, H., Adhikari, R., Mahadevan, I.: A Markov model of the Indus Script. Proc. Nat. Acad. Sci. USA **106**(33) (2009)
10. Revesz, P.Z.: Bioinformatics evolutionary tree algorithms reveal the history of the Cretan Script Family. Inter. J. Appli. Math. Inform. **10**(1), 67–76 (2016)
11. Revesz, P.Z.: Establishing the West-Ugric language family with Minoan, Hattic and Hungarian by a decipherment of Linear A. WSEAS Trans. Inf. Sci. Appl. **14**, 306–335 (2017)
12. Revesz, P.Z.: A vowel harmony testing algorithm to aid in ancient script decipherment. In: Proceedings of the 24th International Conference on Circuits, Systems, Communications and Computers (CSCC), pp. 35–38. IEEE Press (2020)
13. Revesz, P.Z.: Data science applied to discover ancient Minoan-Indus Valley trade routes implied by common weight measures. In: Proceedings of the 26th International Database Engineered Applications Symposium (IDEAS), pp. 150–155. ACM Press (2022), https://doi.org/10.1145/3548785.3548804
14. Revesz, P.Z.: Archaeogenetic data mining supports a Uralic-Minoan homeland in the Danube Basin. Information **15**(10, article number 646) (2024). https://doi.org/10.3390/info15100646
15. Revesz, P.Z.: A tale of two sphinxes: Proof that the Potaissa Sphinx is authentic and other Aegean influences on early Hungarian inscriptions. Mediter. Archaeol. Archaeom. **24**(2), 191–216 (2024). https://doi.org/10.5281/zenodo.13139864
16. Revesz, P.Z.: Trend of increasing percentages of mirror-symmetric signs in the Cretan Script Family and the Phoenician Alphabet Family. In: Understanding Information and its Role as a Tool: In Memory of Mark Burgin, pp. 445–465. World Scientific (2025)
17. VanOrsdale, J., Chauhan, J., Potlapally, S.V., Chanamolu, S., Kasara, S.P.R., Revesz, P.Z.: Measuring vowel harmony within Hungarian, the Indus Valley Script language, Spanish and Turkish using ERGM. In: Proceedings of the 26th International Database Engineered Applications Symposium (IDEAS), pp. 171–174. IEEE Press (2022)
18. Wells, B.K.: The Archaeology and Epigraphy of Indus Writing, Archaeopress (2015)

Author Index

A
Afrane, Mary Dufie 109
Agrawal, Gunjit 252
Alamoudi, Eman 17
Al-Ghezi, Ahmed 211
Alhadidi, Dima 125
Alkhateeb, Banan Mohammad 184
Almousa, Hissah 48
Aslan, Ammar 85

C
Canpolat, Mutlu 85
Çelebi, Selahattin Barış 85

D
Debnath, Chiradeep 252
Devlin, Marie 197

E
Ezhilchelvan, Paul 140

G
Greco, Sergio 96

I
Ishimwe, Joslin 242

K
Kang, Myoung-Ah 171
Kirchner, Sam 34
Kore, Tidenek Fekadu 171

L
Li, Lixin 109
Liu, Ye 140

M
Mason, Michael 34
Meyer, Vanessa 211
Meyer, Volker 71
Mojallal, Kasra 125
Mosleh, Maryam 197

N
Ngiruwonsanga, Prince 242

P
Pinet, François 171
Powell, Carter 34

R
Rabiee, Mehdi 96
Ratwatte, Adrian 242
Revesz, Peter Z. 3, 227, 252

S
Sarramia, David 171
Shaffiey, Shohaib 3
Shahbazian, Reza 96
Siddha, Mohanendra 227
Soares, Edvan 155

Solaiman, Ellis   17, 48, 184, 197
Stahl, Alexander   61

T
Tadi, Ali Abbasi   125
Tamkiya, Harsh   252
Tawadros, Despina   71
Times, Valeria   155
Trubitsyna, Irina   96

W
Wang, Yingming   140
Webber, Jim   140
Wiese, Lena   71, 211

X
Xu, Yao   109

Y
Yang, Wenhui   71

MIX
Papier aus verantwortungsvollen Quellen
Paper from responsible sources
FSC® C105338

If you have any concerns about our products,
you can contact us on
ProductSafety@springernature.com

In case Publisher is established outside the EU,
the EU authorized representative is:
**Springer Nature Customer Service Center GmbH
Europaplatz 3, 69115 Heidelberg, Germany**

Printed by Libri Plureos GmbH
in Hamburg, Germany